Inverse Modeling of the Ocean and Atmosphere

Inverse Modeling of the Ocean and Atmosphere is a graduate-level textbook for students of oceanography and meteorology, and anyone interested in combining computer models and observations of the hydrosphere or solid earth. The scientific emphasis is on the formal testing of models, formulated as rigorous hypotheses about the errors in all the information: dynamics, initial conditions, boundary conditions and data. The products of successful inversions include four-dimensional multivariate analyses or maps of, for example, ocean circulation fields such as temperature, pressure and currents; analyses of residuals in the dynamics, inputs and data; error statistics for all the analyses, and assessments of the instrument arrays or observing systems.

A step-by-step development of maximally-efficient inversion algorithms, using ideal models, is complemented by computer codes and comprehensive details for realistic models. Variational tools and statistical concepts are concisely introduced, and applications to contemporary research models, together with elaborate observing systems, are examined in detail. The book offers a review of the various alternative approaches, and further advanced research topics are discussed.

Derived from the author's lecture notes, this book constitutes an ideal course companion for advanced undergraduates and graduate students, as well as being a valuable reference source for researchers and managers in the theoretical earth sciences, civil engineering, and applied mathematics. Tutors are also directed towards the author's ftp site where they may download complementary overheads for class teaching.

ANDREW BENNETT was awarded a Ph.D. in applied mathematics from Harvard University in 1971. He subsequently continued his research as a National Research Council Fellow at the University of Toronto, and as a Queen's Fellow in Marine Science at Monash University (Melbourne, Australia). Following eight years as a lecturer and senior lecturer in the Department of Mathematics at Monash University, he became a research scientist at the Institute of Ocean Sciences, Sidney, B.C., Canada. He has been a professor at the College of Oceanic and Atmospheric Sciences at Oregon State University since 1987, where his research interests include ocean data assimilation, turbulence theory, and regional modeling. Professor Bennett has won refereeing awards from the *Journal of Physical Oceanography* (1986) and the *Journal of Geophysical Research* (1995), and is also the author of *Inverse Methods in Physical Oceanography* (Cambridge University Press, 1992).

Inverse Modeling of the Ocean and Atmosphere

Andrew F. Bennett
College of Oceanic and Atmospheric Sciences
Oregon State University

CAMBRIDGE
UNIVERSITY PRESS

CAMBRIDGE UNIVERSITY PRESS
Cambridge, New York, Melbourne, Madrid, Cape Town, Singapore, São Paulo

Cambridge University Press
The Edinburgh Building, Cambridge CB2 2RU, UK

Published in the United States of America by Cambridge University Press, New York

www.cambridge.org
Information on this title: www.cambridge.org/9780521813730

First published 2002
This digitally printed first paperback version 2005

A catalogue record for this publication is available from the British Library

Library of Congress Cataloguing in Publication data

Bennett, Andrew F., 1945–
Inverse modeling the ocean and atmosphere / Andrew F. Bennett.
　　p.　cm.
Includes bibliographical references and index.
ISBN 0 521 81373 5
1. Oceanography – Mathematical models.　2. Meteorology – Mathematical models.
3. Inverse problems (Differential equations)　I. Title.
GC10.4.M36 B47　2002
551.46′001′5118 – dc21　　　　　　　　　　　　　　　　　　　2001052655

ISBN-13　978-0-521-81373-0 hardback
ISBN-10　0-521-81373-5 hardback

ISBN-13　978-0-521-02157-9 paperback
ISBN-10　0-521-02157-X paperback

to Elaine, Luke and Antonia

Contents

Preface

Inverse modeling has many applications in oceanography and meteorology. Charts or "analyses" of temperature, pressure, currents, winds and the like are needed for operations and research. The analyses should be based on all our knowledge of the ocean or atmosphere, including both timely observations and the general principles of geophysical fluid dynamics. Analyses may be needed for flow fields that have not been observed, but which are dynamically coupled to observed fields. The data must therefore contribute not only to the analyses of observed fields, but also to the inference of corrections to the dynamical inhomogeneities which determine the coupled fields. These inhomogeneities or inputs are: the forcing, initial values and boundary values, all of which are themselves the products of imperfect interpolations. In addition to input errors, the dynamics will inevitably contain errors owing to misrepresentations of phenomena that cannot be resolved computationally; the data are therefore also required to improve the dynamics by adjusting the empirical coefficients in the parameterizations of the unresolved phenomena. Conversely, the model dynamics must have some credibility, and should be allowed to influence assessments of the effectiveness of observing systems. Finally, and perhaps most compelling of all, geophysical fluid dynamical models need to be formulated and tested as formal scientific hypotheses, so that the development of increasingly realistic models may proceed in an orderly and objective fashion. All of these needs can be met by inverse modeling. The purpose of this book is to introduce recent developments in inverse modeling to oceanographers and meteorologists, and to anyone else who needs to combine data and dynamics.

What, then, is inverse modeling and why is it so named? A conventional modeler formulates and manipulates a set of mathematical elements. For an ocean model, the set includes at least the following:

(i) a domain in four-dimensional space, representing an ocean region and a time interval of interest;

(ii) a system of inhomogeneous partial differential equations expressing the phenomenological dynamics of the circulation (there will be inhomogeneities owing to internal fields which force the dynamics beneath the ocean surface, and the equations will include empirical parameters representing unresolved phenomena);

(iii) initial conditions for the equations, representing the ocean circulation or state at some time; and

(iv) boundary conditions which may be inhomogeneous owing either to forcing of the ocean at the ocean surface, or to fields of flow and thermodynamic conditions imposed at lateral open boundaries.

There will be subtle yet profoundly important differences for, say, atmospheric models; consider the character of boundary conditions, for example. However, ocean models will be invoked henceforth as the default choice for concise discussion.

From a mathematical perspective, the partial differential operators, initial operators and boundary operators can all be seen as acting in combination upon the solution for the ocean circulation, and producing the inhomogeneities or inputs which are, again, the subsurface forcing, initial values, surface forcing and any values at open boundaries. The combined operator is nonsingular if there exists a unique and analytically satisfactory – say, continuously differentiable – solution for each set of analytically satisfactory inputs. If the operator is nonsingular, then there is a well-defined and unique inverse operator. From the mathematical perspective, the solution is the action of the inverse operator on the inputs. Computing this action is, in the infinite wisdom of convention, "forward ocean modeling".

Characterizing our knowledge of ocean circulation as solutions of well-posed, mixed initial-value boundary-value problems does not correspond to our real experience of the ocean. Ship surveys, moored instruments, buoys drifting freely on the ocean surface or floating freely below the surface, and earth satellites orbiting above cannot observe continuous fields throughout an ocean region, even for one instant. Yet these data, after control for quality, are in general far more reliable than either the parameterizations of turbulence in the dynamics or the crudely interpolated forcing fields, initial values and boundary values. The quality-controlled data belong to any rational concept of a model, and the set of mathematical elements that defines a model is readily extended to include them. Specifically, functionals corresponding to methods of measurement, and numbers corresponding to measurements of quantities in the real ocean (such as velocity components, temperature, density and the like), may be added to the set. Each functional maps a circulation field into a single number. For example, monthly-mean sea level at a coastal station defines a kernel or integrand which selects sea level from the many circulation variables, which has a rectangular time window of one month and which is sharply peaked at the coastal station. Note that the new mathematical elements include both additional operators (the measurement functionals), and additional input (the data).

It is always assumed, but almost never proved, that the operator for the original model is nonsingular. It can always be assumed that applying the measurement functionals to the unique solution of the original forced, initial-boundary value problem does not produce numbers equal to the real data. The extended operator can therefore have no inverse, and so must be singular. It seems natural, even a compulsion (Reid, 1968), to determine the ocean circulation as some uniquely-defined best-fit to the extended inputs (forcing, initial, boundary and observed). The singular extended operator then has a generalized inverse operator, and the best-fit ocean circulation is the action of the generalized inverse on the extended inputs. This book outlines the theoretical and practical computation of the action, for best-fits in the sense of weighted least-squares. The practical computations will be only numerical approximations, so the theme of the book should therefore be expressed as "inverting numerical models and observations of the ocean and atmosphere in a generalized sense". Abandoning precision for brevity, the theme is "inverse modeling the ocean and atmosphere".

How, then, does inverse modeling meet the needs of oceanographers and meteorologists? The best-fit circulation is clearly an analysis, an optimal dynamical interpolation in fact, of the observations. All the fields coupled by the dynamics are analyzed, even if only some of them are observed. The least-squares fit to all the information, observational and dynamical, yields residuals in the equations of motion as well as in the data, and these residuals may be interpreted as inferred corrections to the dynamics or to the inputs. There are emerging techniques that can in principle distinguish between additive errors in dynamics and internal forcing, but these techniques are so new and unproven that it would be premature, even by the standards of this infant discipline, to include them here. Empirical parameters may also be tuned to improve the analysis. (The tuning game, sometimes described as a "fiddler's paradise" [Ljung and Söderström, 1987], is outlined here.) The conditioning or sensitivity of the fit to the inputs, as revealed during the construction of the generalized inverse, quantifies the effectiveness of the observing system. The natural choices for the weights in the best fit are inverses of the covariances of the errors in all the operators and inputs. These covariances must be stipulated by the inverse modeler. They accordingly constitute, along with stipulated means, a formal hypothesis about the errors in the model and observations. The minimized value of the fitting criterion or penalty functional yields a significance test of that hypothesis. For linear least-squares, the minimal value is the χ^2 variable with as many degrees of freedom as there are data, provided the hypothesized means and covariances are correct. A failed significance test does discredit the analyzed circulation and also any concomitant assessment of the observing system, but does not end the investigation: detailed examination of the residuals in the equations, initial conditions, boundary conditions and data can identify defects in the model or in the observing system. Thus model development can proceed in an orderly and objective fashion. This is not to deny the crucial roles of astute and inspired insight in oceanic and atmospheric model development as in all of science; it is rather to advocate a minimal level of organization especially when inspiration is failing us, as seems to be the case at present.

In spite of an explicit emphasis here on time dependence, the spirit of this approach is close to geophysical inverse theory (see for example Parker, 1994), specifically the estimation of permanent strata in the solid earth using seismic data. The retrieval of instantaneous vertical profiles of atmospheric temperature and moisture using multi-channel microwave soundings from satellites (see for example Rodgers, 2000) bears a striking resemblance to the seismic problem, and is indeed both named and practised as inverse theory. Yet the context here – time-dependent oceanic and atmospheric circulation – is so different that to call it inverse theory seems almost misleading.

Inverse modeling is but one formulation of the vaguely defined activity known as "data assimilation". The most widely practised form of oceanic or atmospheric data assimilation involves interpolating fields at one time, for subsequent use as initial data in a model integration which may even be a genuine forecast. Once nature has caught up with the forecast, the latter serves as a first-guess or "background" field for the next synoptic analysis. As might be imagined, this cycle of synoptic analysis and fore-casting is a major enterprise at operational centers, and is very extensively developed for meteorological applications. Characterization of operational systems for observing the weather, in particular studying the statistics of observational errors, has been and remains the subject of vast investigation. Comprehensive references may be found at appropriate places in the following chapters, but that description of operational detail will not be repeated here. Nor will the intricate, "diagnostically-constrained" multivari-ate forms of synoptic interpolation be discussed in detail. Geostrophy, for example, is an approximate diagnostic constraint on synoptic fields of velocity and pressure. The emphasis instead will be on elaborating the new data assimilation schemes that could be consistently described as nonsynoptic, "prognostically-constrained" interpolation. The unapproximated law of conservation of momentum, for example, is a prognos-tic constraint. Again, the nature of this latter activity is so different in technique and broader in scope, in comparison with the conventional cycle of synoptic analysis and forecasting, that to call these new schemes "data assimilation" seems to be misleading yet again. The name "inverse modeling" is chosen, for better or worse.

What else has been left out here? Monte Carlo methods are immensely appealing in any application, and data assimilation is no exception. Sample estimates of means and covariances of circulation fields may be generated from repeated forward integra-tions of a model driven by suitably constructed pseudo-random inputs. The sample moments of the solutions are then used for conventional synoptic interpolation. The calculus of variations is not required. These assimilation methods, especially "ensem-ble Kalman filtering", are highly competitive with variational inverse methods in terms of development effort and computational efficiency, but are even more immature and so are mentioned only briefly. The very basics of statistical simulation and Monte Carlo methods in general are outlined in these chapters.

The reader should not be discouraged by the technical definition of inverse modeling given in previous paragraphs. The calculus of several variables, a rudimentary knowledge of partial differential equations and the same numerical analysis used to solve the forward model are enough mathematics for the computation of generalized inverses. Abstraction is restricted to the one place in this book where an elegant expression of generality is of real benefit. The Hilbert Space analysis sketched in Chapter 2 exposes the geometrical structure of the generalized inverse, and explains the efficiency of the concrete algorithms developed in Chapter 1. The geometrical interpretation is a straightforward adaptation of the theory of Laplacian spline interpolation. A beautiful treatment of L-splines may be found in an applied meteorology journal (Wahba and Wendelberger, 1980), to the eternal credit of the authors, reviewers and editors. Any temptation to make use of the Hilbert Space machinery for abstract definitions of adjoint operators is easily resisted, as such abstraction offers no real insight into the problem of interest. The adjoint operators arise naturally when the elementary calculus of variations is used to derive the classic Euler–Lagrange conditions for the weighted, least-squares best-fit. Unlike the Hilbert Space definition of an adjoint operator, the variational calculus need not be preceded by a linearization of the dynamics and measurement functionals. This flexibility leads to critically important alternatives for iterative solution techniques that are linear.

It is essential to distinguish the formulative and interpretive aspects of inverse modeling from its mathematical aspects. Least-squares may be used to estimate any quantity, but it is the estimator of maximum likelihood for Gaussian or normal random variables. Such variability can reasonably be expected in the ocean and atmosphere, on the synoptic scale and larger, away from transient and semi-permanent fronts, and in variables not subject to phase changes. Least-squares is especially attractive from a mathematical perspective, since it leads to linear conditions for the best fit when the constraints are also linear. The linearity of the extremal conditions permits powerful analyses which yield efficient solution methods. There are many least-squares algorithms, such as optimal interpolation, Kalman filtering, fixed-interval smoothing, and representers. The relationships between these statistical, control-theoretic and geometrical approaches are explained in this book. Aside from unifying the mathematics, recognizing the mathematical relationships facilitates the identification of scientific assumptions.

For example, if the data were collected in much less time than the natural scales of evolution of the dynamics and the internal forcing, then there would be little to gain by assuming that there are errors in the dynamics or internal forcing. It would suffice to admit errors only in the initial conditions, surface forcings, open boundary values and data. This assumption massively reduces the finite dimension of the "state space" for the numerical model, by eliminating those variable fields or "controls" defined both throughout the ocean region and throughout the time interval of interest. Boundary values, initial values and empirical parameters would be retained as controls. The reduced state or "control" space may be sufficiently small that a conventional gradient search for a minimum in the state space is feasible. The condition of the fit in state space determines the efficiency of the search. It is, however, becoming increasingly necessary to consider time intervals during which the dynamical errors are bound to become significant. That is, the initial conditions would be ineffective as controls for guiding the model solution towards the later data. Using distributed controls, that is, admitting errors in the dynamics throughout space and time, leads to huge numbers of computational degrees of freedom. (There are in general far fewer statistically independent degrees of freedom, but these are not readily identified. Indeed, the methods developed here serve to identify them.) Hence there could be no prospect of a well-conditioned search in the control space or in the equivalent state space. The power of the methods described in these chapters is that they identify a huge subspace of controls (known as the null space) having exactly no influence on guiding the solution towards the data. The methods restrict the search for optimal controls to those lying entirely in the comparatively tiny, orthogonal complement of the null space (known as the data subspace). Again, as in the choice of a least-squares estimator, there is an interplay between scientific formulation and mathematical technique. The two should nonetheless always be carefully distinguished.

As a final example of the distinction and interplay between scientific formulation and mathematical manipulation, consider errors in models of small-scale flows. As already implied, these errors are likely to be highly intermittent or nonGaussian. Thus, inversions of observations collected in mixing fronts and jets, or in free convection, or during phase changes, will require estimators other than least squares. Only brute-force minimization techniques, such as simulated annealing or Monte Carlo methods in general, appear to be available for most estimators. On the other hand, multi-processor computer architecture may favor brute-force inversion. These brute-force techniques will be mentioned here, but only briefly, since by their nature regrettably little is known about them.

The content of this book closely follows an upper-level graduate course for physical oceanography students at Oregon State University. Their preparation includes

- graduate courses in fluid dynamics, geophysical fluid dynamics and ocean circulation theory;
- a graduate course in numerical modeling of ocean circulation;
- a graduate course in time series analysis including Gauss–Markov smoothing or "objective analysis";
- graduate courses in ordinary and partial differential equations, computational linear algebra and numerical methods in general;
- FORTRAN and basic UNIX skills;
- or an equivalent preparation in atmospheric science.

The curriculum does not require great depth or fresh familiarity with all of the above material. The following would suffice.

1. Some minimal exposure to hydrodynamics, preferably in a rotating reference frame, including approximations such as hydrostatic balance, the shallow-water equations and geostrophic balance. The well-known texts by Batchelor (1973), Pedlosky (1987), Gill (1982), Holton (1992) and Kundu (1990) may be consulted. Graduate students in physics or mechanical or civil engineering would have no problem with the curriculum, although some jargon may cause them to glance at a text in oceanography or meteorology.

2. The knowledge that oceanic and atmospheric circulation models are expressed as partial differential equations (pdes) that may be numerically integrated, most simply using finite differences. The text by Haltiner and Williams (1980) on numerical weather prediction is very useful.

3. Access to Stakgold's classic (1979) text on boundary value problems. The theoretical notions most useful here are (i) odes and pdes can only have well-behaved solutions if precisely the right number of initial and boundary value conditions are provided and (ii) the solution of such well-posed problems for linear odes and pdes can be expressed using a Green's function or influence function. As for computational linear algebra and numerical methods in general, the synopses in Press et al. (1986) are very useful.

4. Comfort with the very basics of probability and statistics, including random variables, means, covariances and minimum-variance estimation. Again, the synopses in Press et al. (1986) make a good first reading.

5. As much FORTRAN as can be learned in a weekend.

The content of the Preamble, and of each of the six chapters and the two appendices, is outlined on their first pages. The Preamble attempts to communicate the nature of variational ocean data assimilation, or any other assimilation methodology, through a commonplace application of basic scientific method to marine biology. The example might seem out of context, and indeed it is, but that underscores the universality and long history of the approach advocated here. Its arrival in the context of oceanic and atmospheric circulation has of course been delayed by the fantastic mathematical and computational complexity of circulation models. The Preamble includes a "data assimilation checklist", which the student or researcher is encouraged to consult regularly. Chapter 1 is the irreducible introduction to variational assimilation with dynamical models; a "toy" model consisting of a single linear wave equation with one space dimension serves as an illustration. Chapter 2 complements the control-theoretic development of Chapter 1 with geometrical and statistical interpretations; analytical considerations essential to the physical realism of the inverse solutions are introduced. Chapter 3 addresses efficient construction of the inverse and its error statistics, and introduces iterative techniques for coping with nonlinearity. Chapter 4 surveys alternative algorithms for linear least-squares assimilation, and for assimilation with nonlinear or nonsmooth models or with nonlinear measurement functionals. Difficulties to be expected with nonlinear techniques are outlined – proven remedies are still lacking. Chapter 5 reviews large-scale geophysical fluid dynamics, discusses several real oceanic and atmospheric inverse models in detail, and concludes with notes on a selection of contemporary efforts, both research and operational. Chapter 6 applies inverse methods to forward models based on singular operators.

The material in this book can be presented in thirty one-hour lectures. An overhead projector is a great help: minimal-text, math-only, large-font overhead transparencies allow the audience to listen, rather than transcribe incorrectly. The overheads are available as TEX source files via an anonymous ftp site (`ftp.oce.orst.edu`, `dist/bennett/class/overheads`). Students should be able to begin the computing exercises in Appendix A after studying the first four sections of Chapter 1. The inverse tidal model of §5.2 in Chapter 5 is accessible after studying Chapters 1 and 2. The nonlinear inverse models of tropical cyclones and ENSO in §5.3–5.5, and the accelerated algorithms used in their construction, require a study of Chapter 3. The complete variational equations for the tropical cyclone inversion may be found in Appendix B. A first reading of Chapter 4 is assumed in §5.6, the survey of contemporary applications of advanced assimilation with oceanic and atmospheric data.

The research monograph by Bennett (1992) contains almost all of the theoretical development found here, but none of the guidelines for implementation and few case studies with real data or real arrays. Certain advanced theoretical considerations, such as Kalman filter pathology in the equilibrium limit and continuous families of representers for excess boundary data, are only briefly mentioned here if at all, but may be found in the earlier monograph. There has been a rapid growth in the literature of nonsynoptic data assimilation during the last decade. A full literature survey would be impractical and of doubtful value as so much work has been highly application-specific. Shorter but very useful survey articles include Courtier *et al.* (1993); Anderson, Sheinbaum and Haines (1996) and Fukumori (2001); for collections of expository papers and applications see Malanotte-Rizzoli (1996), Ghil *et al.* (1997) and Kasibhatla *et al.* (2000). The last-mentioned is noteworthy for its interdisciplinary range, and also for a set of exercises on various assimilation techniques. The major text by Wunsch (1996) principally develops in great detail the time-independent inverse theory for steady ocean circulation, using a finite-dimensional formulation which certainly complements the analytical development here and which may be the more accessible for being finite-dimensional. On the other hand the essential mathematical condition of the inverse problem is established at the analytical or continuum level, and the "look and feel" of geophysical fluid dynamics is retained by an analytical formulation.

Inverse modeling suffers not so much from the lack of good data, credible models and adequate computing resources, as from a lack of experience. This book is intended to be of assistance to the generation of investigators who, it is hoped, will acquire that experience.

Monterey, June 2001

Acknowledgements

Many of my collaborators and colleagues over the last two decades have influenced this book, most recently Boon Chua, Gary Egbert and Robert Miller. The exercises in Appendix A were devised and constructed by Boon Chua and Hans-Emmanuel Ngodock. The book is based upon lecture notes for two summer schools on inverse methods and data assimilation, held at Oregon State University in 1997 and 1999. Numerous suggestions from the summer school participants led to significant improvement of the notes. Successive versions were all patiently typed by Florence Beyer, and the book manuscript was created from the notes by William McMechan.

Permission to reproduce copyrighted figures was received from: Inter-Research (Fig. P.1.1), Elsevier (Fig. 5.2.7), the American Geophysical Union (Fig. 5.2.8), Nature (Fig. 5.3.5), Springer-Verlag (Figs. 5.4.1, 5.4.2, 5.4.3, 5.4.5), and the American Meteorological Society (Figs. 5.5.1, 5.5.2, 5.5.3, 5.5.4, 5.5.5, 5.5.6, 5.5.7, 5.5.8). Dudley Chelton kindly provided Fig. 5.2.5. All the other figures were drafted by David Reinert. Thanks for the cover illustrations are owed to Michael McPhaden and Linda Stratton of the NOAA Pacific Marine Environmental Laboratory. The support of the National Science Foundation (OCE-9520956) is also acknowledged.

It has been a pleasure to work with Cambridge University Press: my editors Alan Harvey and Matt Lloyd; also Nicola Stern, Susan Francis, Jo Clegg, and Frances Nex.

The manuscript was completed at the Office of Naval Research Science Unit in the Fleet Numerical Meteorology and Oceanography Center, under the auspices of the Visiting Scientist Program of the University Corporation for Atmospheric Research. I am grateful to Oregon State University for an extended leave of absence, to Manuel

Fiadeiro at ONR and Michael Clancy at FNMOC, to Meg Austin and her team at the UCAR VSP Office for their friendly efficiency, and to Captain Joseph Swaykos, USN and his entire staff, for their hospitality during an educational and productive stay at "Fleet Numerical".

Internet sites mentioned in this book:

bioloc.coas.oregonstate.edu/pictures/gallery2/index.html

ftp.oce.orst.edu, cd/dist/chua/IOM/IOSU, 216

ftp.oce.orst.edu, cd/dist/bennett/class, 146, 195

ftp.oce.orst.edu, cd/dist/bennett/class/overheads, xx

www.gfdl.gov/šmg/MOM/MOM.html, 167

www.polar.gsfc.nasa.gov 166

www.fnmoc.navy.mil, 166

www.pac.dfo-mpo.gc.ca/sci/osap/projects/plankton/zoolab_e.htm#Copepod

www.oce.orst.edu/po/research/tide/global.html, 136

www.pmel.noaa.gov/toga-tao/home.html, 156

http://diadem.nersc.no/project, 169

www.units.it/m̃abiolab/set_previous.htm, 1

Note: oce.orst.edu changes to coas.oregonstate.edu in 2002.

The publisher has used its best endeavors to ensure that the URLs for external websites referred to in this book are correct and active at the time of going to press. However, the publisher has no responsibility for the websites and can make no guarantee that a site will remain live or that the content is or will remain appropriate.

Preamble

An ocean data assimilation system in miniature

The pages of this book are filled with the mathematics of oceanic and atmospheric circulation models, observing systems and variational calculus. It would only be natural to ask: What is going on here, and is it really new? The answers are "regression" and hence "no": almost every issue of any marine biology journal contains a variational ocean data assimilation system in miniature.

P.1   Linear regression in marine biology

The article "Repression of fecundity in the neritic copepod *Acartia clausi* exposed to the toxic dinoflagellate *Alexandrium lusitanicum*: relationship between feeding and egg production", by Jörg Dutz, appeared in *Marine Ecology Progress Series* in 1998. Dinoflagellates are a species of phytoplankton, or small plant-like creatures. The genus *Alexandrium* (www.units.it/m̃abiolab/set_previous.htm, click on 'Toxic microalgae') produces toxins which rise through the food web to produce paralytic shellfish poisoning in a variety of hydrographical regions, ranging from temperate to tropical. Zooplankton, or small animal-like creatures (www.ios.bc.ca/ios/plankton/ios_tour/zoop_lab/copepod.htm), graze on these dinoflagellates. The effect of the toxins on the grazers naturally arises. Dutz (1998) fed toxin-bearing *Alexandrium lusitanicum* and toxin-free *Rhodomonas baltica* (bioloc.coas.oregonstate.edu/baltica.jpg) to females of the copepod *Acartia clausi* in controlled amounts, and measured the fecundity or gross growth

1

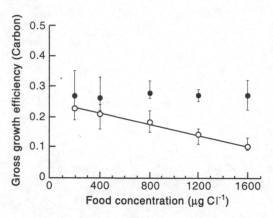

Figure P.1.1 Gross growth efficiency of *Acartia clausi* versus food supply. Solid circles: nontoxic *Rhodomonas baltica*; open circles: toxic *Alexandrium lusitanicum* (after Dutz, 1998).

efficiency in terms of total carbon production. He found that the grazers were not killed, and they continued to lay eggs. However, their fecundity was affected: see Fig. P.1.1. Note the controlled food concentration (abscissa x) with five values: 200, 400, 800, 1200, and 1600 μg Cl^{-1}. Fecundity is not influenced by the supply of nontoxic *Rhodomonas* (solid circles), but is clearly reduced as the supply of toxic *Alexandrium* (open circles) increases. The gross growth efficiencies (ordinate y) in the latter case are respectively: 0.23, 0.21, 0.18, 0.14, 0.10 (Dutz, 1998; Table 2). The error bars indicate Dutz' maximum and minimum estimates. A straight line clearly fits the *Alexandrium* data well. The regression parameters are: $a = 0.25$, $b = 9.2 \times 10^{-5}$, $r^2 = 0.997$, $F_{1,3} = 355$, $P < 0.0005$.

A brief review of linear regression is in order. The data are M ordered pairs: (x_m, y_m), $1 \le m \le M$. The model is

$$y_m = \alpha + \beta x_m + \epsilon_m, \tag{P.1.1}$$

where α and β are unknown constants, while ϵ_m is a random variable with mean and covariance

$$E\epsilon_m = 0, \quad E(\epsilon_m \, \epsilon_n) = \sigma^2 \delta_{nm} = \begin{cases} \sigma^2, & n = m \\ 0, & n \ne m. \end{cases} \tag{P.1.2}$$

The error ϵ_m is an admission of measurement error, and of the unrepresentativeness of a linear relationship. Note that the model consists of an explicit functional form (here, a linear relationship), together with probabilistic statements (here, mean and covariance) about the error in the form. We seek an estimate (here, a regression line):

$$\hat{y} = a + bx, \tag{P.1.3}$$

where a and b are to be chosen. As an estimator, let us choose a uniformly weighted sum of squared errors:

$$WSSE = \sigma^{-2} \sum_{m=1}^{M} (y_m - a - bx_m)^2. \tag{P.1.4}$$

A value for σ may be inferred from the error bars in Fig. P.1.1. It is easily shown that $WSSE$ is minimal if a and b satisfy the normal equations:

$$\begin{pmatrix} 1 & \bar{x} \\ \bar{x} & \overline{x^2} \end{pmatrix} \begin{pmatrix} a \\ b \end{pmatrix} = \begin{pmatrix} \bar{y} \\ \overline{xy} \end{pmatrix}, \tag{P.1.5}$$

where the overbar denotes the arithmetic mean, for example $\bar{x} = M^{-1} \sum_{m=1}^{M} x_m$. Note that (P.1.5) is independent of the uniform weight σ^{-2}. These equations are of course trivially solved for a and b. The following statements may be made about the first and second moments of the solution:

$$Ea = \alpha, \quad Eb = \beta,$$

$$E(a - \alpha)^2 = \frac{\overline{x^2} \sigma^2}{M(\overline{x^2} - (\bar{x})^2)}, \quad E(b - \beta)^2 = \frac{\sigma^2}{M(\overline{x^2} - (\bar{x})^2)}. \tag{P.1.6}$$

Moreover, a, b and $\hat{y}_m$ are normally distributed around α, β and y_m respectively. Note that the error variances in (P.1.6) are $O(M^{-1})$. In addition to the posterior error estimates (P.1.6), there are significance test statistics such as the variance-ratio or F test:

$$F_{1,M-2} = \frac{\sum_{m=1}^{M} (y_m - \bar{y})^2}{\sum_{m-1}^{M} (y_m - \hat{y}_m)}, \tag{P.1.7}$$

where $\hat{y}_m \equiv ax_m + b$. The numerator is the total variance of the data; the denominator is the total variance of the residuals for the regression line (P.1.3). Note that (P.1.7) is independent of σ^2. The subscripts 1 and $M - 2$ indicate the number of degrees of freedom in the denominator and the numerator, respectively. The value of F here is 355; accordingly the probability P of the null hypothesis ($\alpha = \beta = 0$) being true is less than 0.05%. In other words it is highly credible that grazing on *Alexandrium lusitanicum* does repress the fecundity of *Acartia clausi*.

Exercise P.1.1
An alternative test statistic is provided by the weighted denominator in (P.1.7):

$$resWSSE = \sigma^{-2} \sum_{m=1}^{M} (y_m - \hat{y}_m)^2$$

$$\sim \chi_M^2, \quad \text{as} \quad M \to \infty. \tag{P.1.8}$$

Verify that $E\chi_M^2 = M$, $\text{var}\chi_M^2 = 2M$. Calculate (P.1.8) using Dutz' data, and draw conclusions. $\square$

If the data had suggested it, Dutz could have considered quadratic regression:

$$y_m = \alpha + \beta x_m + \gamma x_m^2 + \epsilon_m,$$

$$E\epsilon_m = 0, \quad E(\epsilon_n \epsilon_m) = \sigma^2 \delta_{nm}. \tag{P.1.9}$$

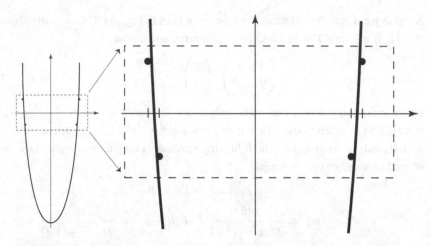

Figure P.1.2 On the left: the parabola of least-squares best fit to four data points, which are shown as solid circles. The abscissa values for the data (see the tick marks on the abscissa in the zoom on the right) are ill-chosen. As a result, the least-squares best fit is clearly ill-conditioned. The abscissa itself would be a more sensible fit to the data.

The estimate would be

$$\hat{y} = a + bx + cx^2. \tag{P.1.10}$$

The estimator would again be (P.1.4), for which the normal equations are

$$\begin{pmatrix} 1 & \overline{x} & \overline{x^2} \\ \overline{x} & \overline{x^2} & \overline{x^3} \\ \overline{x^2} & \overline{x^3} & \overline{x^4} \end{pmatrix} \begin{pmatrix} a \\ b \\ c \end{pmatrix} = \begin{pmatrix} \overline{y} \\ \overline{xy} \\ \overline{x^2 y} \end{pmatrix}. \tag{P.1.11}$$

Suppose for simplicity that $\overline{x} = \overline{x^3} = 0$ (these are at our disposal). Then the system (P.1.11) is ill-conditioned; that is, the solution (a, b, c) is highly sensitive to the inhomogenity on the right-hand side if $\overline{x^4}/(\overline{x^2})^2 \ll 1$. This ratio is also at our disposal. Just such a situation is sketched in Fig. P.1.2. The best fit to the four data points is a deep parabola, yet the most sensible fit would be the abscissa itself ($y = 0$). In conclusion, the stability of the estimate (P.1.10) is controlled by the choice of abscissa values x_m, $1 \le m \le M$.

P.2   Data assimilation checklist

The preceeding elementary application of linear regression in marine biology has every aspect of an "ocean data assimilation system": see the following checklist.

Data assimilation checklist

INPUTS
(i) **There is an observing system**, consisting of measurements of gross growth efficiency at selected food concentration levels.

(ii) **There are dynamics**, expressed here as (P.1.1), the explicit general solution of the differential equation

$$\frac{d^2y}{dx^2} = 0, \tag{P.2.1}$$

plus measurement errors ϵ_m, $1 \le m \le M$. The values α, β indicated in (P.1.1) for the regression constants a, b are the "true" values.

(iii) **There is an hypothesis** (P.1.2) about the distribution of errors ϵ_m around the true regression line.

(iv) **There is an estimator**, here the uniformly weighted sum of squared errors (P.1.4).

(v) **There is an optimization algorithm**, here the normal equations (P.1.5) which would, in the general case of Nth-order polynomial regression, be robustly solved using the singular value decomposition.

OUTPUTS

(vi) **There is an estimate of the state**, here the regression line (P.1.3) with values of a and b obtained from the normal equations (P.1.5).

(vii) **There are estimates of data residuals and dynamical residuals.** Here the two types of residual are indistinguishable; both are in fact given by $y_m - \hat{y}_m$.

(viii) **There are posterior error statistics**, here the means and variances (P.1.6) for $a - \alpha$ and $b - \beta$.

(ix) **There is an assessment of the array or observing system.** Here it is the conditioning of the normal matrix, and is determined by the choices of food concentrations x_m, $1 \le m \le M$.

(x) **There are test statistics**, here the F-variable (P.1.7) and χ^2-variable (P.1.8). These indicate the credibility of the hypothetical model, and thus the credibility of the derived posterior error statistics.

(xi) **There are indications for model improvement.** Here, however, the indication is that the linear model is so credible that a quadratic model (P.1.10) is unnecessary.

Variational assimilation of El Niño data from the tropical Pacific, into a coupled intermediate model of the ocean and atmosphere, is described in §5.5. The checklist reads as follows.

INPUTS

(i) The observations are monthly-mean and five-day mean values of Sea Surface Temperature (SST, or $T^{(1)}$), the depth of the 20° isotherm ($Z20$) and surface winds (u^a, v^a), at the TOGA–TAO moorings, from April 1994 to May 1998.

(ii) The dynamics are those of an intermediate coupled model after Zebiak and Cane (1987); the thermodynamics of the upper oceanic layer and the coupling through the wind stress are nonlinear. Otherwise the oceanic and atmospheric dynamics are those of linearized shallow-water waves.

(iii) The hypothesis consists of means and autocovariances of errors in the dynamics, in the initial conditions and in the data.
(iv) The estimator is the combined, space-integrated and time-integrated weighted squared error.
 (v) The optimization algorithm is the iterated, indirect representer algorithm for solving the nonlinear Euler–Lagrange equations.

OUTPUTS

 (vi) There are estimates of space-time fields of surface temperature, currents, thermocline depths and surface winds.
(vii) There are corresponding space–time fields of minimal residuals in the dynamics, initial conditions and data.
(viii) There are space–time covariances of errors in the optimal estimates of the coupled circulation.
 (ix) These are assessments of the efficiency of the monthly-mean TOGA–TAO system for observing the "weak" dynamics of the coupled model, that is, observing the intermediate dynamics subject to the hypothesized error statistics.
 (x) The reduced estimator is a χ^2-variable for testing the hypothesized error moments (they were found to lack credibility).
 (xi) The dominance of the minimal residual in the upper-ocean thermodynamic balance indicates that it would serve no purpose to hypothesize increased variances for the dynamical errors: the low-resolution intermediate dynamics should be abandoned in favor of a fully-stratified, high-resolution, Primitive Equation model.

Variational data assimilation, or generalized inversion of dynamical models and observations, is really no more than regression analysis. The novelty lies in the mathematical and physical subtlety of realistic dynamics, in the complexity of the hypotheses about the multivariate random error fields, and in the sheer size of modern data sets. The novelty also lies in the emergence of powerful and efficient optimization algorithms, which allow us to test our models in the same way that all other scientists test theirs.

Chapter 1

Variational assimilation

Chapter 1 is a minimal course on assimilating data into models using the calculus of variations. The theory is introduced with a "toy" model in the form of a single linear partial differential equation of first order. The independent variables are a spatial coordinate, and time. The well-posedness of the mixed initial-boundary value problem or "forward model" is established, and the solution is expressed explicitly with the Green's function. The introduction of additional data renders the problem ill-posed. This difficulty is resolved by seeking a weighted least-squares best fit to all the information. The fitting criterion is a penalty functional that is quadratic in all the misfits to the various pieces of information, integrated over space and time as appropriate. The best-fit or "generalized inverse" is expressed explicitly with the representers for the penalty functional, and with the Green's function for the forward model. The behavior of the generalized inverse is examined for various limiting choices of weights. The smoothness of the inverse is seen to depend upon the nature of the weights, which will be subsequently identified as kernel inverses of error covariances. After reading Chapter 1, it is possible to carry out the first four computing exercises in Appendix A.

1.1   Forward models

1.1.1   Well-posed problems

Mechanics is captured mathematically by "well-posed problems". The mechanical laws for particles, rigid bodies and fields are with few exceptions expressed as ordinary or partial differential equations; data about the state of the mechanical system are provided in initial conditions or boundary conditions or both. The collection of general equations and ancillary conditions constitute a "well-posed problem" if, according to Hadamard (1952; Book I) or Courant and Hilbert (1962; Ch. III, §6):

(i) a solution exists,

which

(ii) is uniquely determined by the inputs (forcing, initial conditions, boundary conditions),

and which

(iii) depends continuously upon the inputs.

Classical particles and bodies move smoothly, while classical fields vary smoothly so only differentiable functions qualify as solutions. The repeatability of classical mechanics argues for determinism. The classical perception of only finite changes in a finite time argues for continuous dependence.

Ill-posed problems fail to satisfy at least one of conditions (i)–(iii). They cannot be solved satisfactorily but can be resolved by generalized inversion, which is the subject of this chapter. Inevitably, well-posed problems are also known as "forward models": given the dynamics (the mechanical laws) and the inputs (any initial values, boundary values or sources), find the state of the system. In this first chapter, an example of a forward model is given; the uniqueness of solutions is proved, and an explicit solution is constructed using the Green's function. That is, the well-posedness of the forward model is established.

1.1.2   A "toy" example

The following "toy" example involves an unknown "ocean circulation" $u = u(x, t)$, where x, t and u are real variables. The "ocean basin" is the interval $0 \le x \le L$, while the time of interest is $0 \le t \le T$: see Fig. 1.1.1.

The "ocean dynamics" are expressed as a linear, first-order partial differential equation:

$$\frac{\partial u}{\partial t} + c\frac{\partial u}{\partial x} = F \qquad (1.1.1)$$

for $0 \le x \le L$ and $0 \le t \le T$, where c is a known, constant, positive phase speed. The inhomogeneity $F = F(x, t)$ is a specified forcing field; later it will become known as the prior estimate of the forcing. An initial condition is

$$u(x, 0) = I(x) \qquad (1.1.2)$$

for $0 \le x \le L$, where I is specified. A boundary condition is

$$u(0, t) = B(t) \qquad (1.1.3)$$

for $0 \le t \le T$, where B is specified.

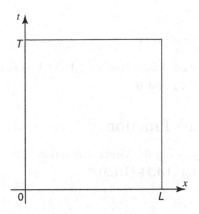

Figure 1.1.1 Toy ocean basin.

1.1.3 Uniqueness of solutions

To determine the uniqueness of solutions (Courant and Hilbert, 1962) for (1.1.1), (1.1.2) and (1.1.3), let u_1 and u_2 be two solutions for the same choices of F, I and B. Define the difference

$$v \equiv u_1 - u_2. \tag{1.1.4}$$

Then

$$\frac{\partial v}{\partial t} + c \frac{\partial v}{\partial x} = 0 \tag{1.1.5}$$

for $0 \leq x \leq L$ and $0 \leq t \leq T$;

$$v(x, 0) = 0 \tag{1.1.6}$$

for $0 \leq x \leq L$, and

$$v(0, t) = 0 \tag{1.1.7}$$

for $0 \leq t \leq T$.

Multiplying (1.1.5) by v and integrating over all x yields

$$\frac{d}{dt} \frac{1}{2} \int_0^L v^2 \, dx = -c \left[\frac{1}{2} v^2 \right]_{x=0}^{x=L} = -\frac{c}{2} v(L, t)^2, \tag{1.1.8}$$

using the boundary condition (1.1.7). Integrating (1.1.8) over time from 0 to t yields

$$\frac{1}{2} \int_0^L v^2(x, t) \, dx = \frac{1}{2} \int_0^L v^2(x, 0) \, dx - \frac{c}{2} \int_0^t v^2(L, s) \, ds. \tag{1.1.9}$$

The right-hand side (rhs) of (1.1.9) is nonpositive, as a consequence of the initial condition (1.1.6). Hence

$$v(x, t) = 0, \tag{1.1.10}$$

that is,

$$u_1(x, t) = u_2(x, t) \tag{1.1.11}$$

for $0 \leq x \leq L$ and $0 \leq t \leq T$. So we have established that (1.1.1), (1.1.2) and (1.1.3) have a unique solution for each choice of F, I and B.

1.1.4   Explicit solutions: Green's functions

We may construct the solution explicitly, using the Green's function (Courant and Hilbert, 1953) or fundamental solution γ for (1.1.1)–(1.1.3).

Let $\gamma = \gamma(x, t, \xi, \tau)$ satisfy

$$-\frac{\partial \gamma}{\partial t} - c \frac{\partial \gamma}{\partial x} = \delta(x - \xi)\delta(t - \tau), \tag{1.1.12}$$

where the δs are Dirac delta functions, and $0 \leq \xi \leq L, 0 \leq \tau \leq T$. Also,

$$\gamma(L, t, \xi, \tau) = 0 \tag{1.1.13}$$

for $0 \leq t \leq T$, and

$$\gamma(x, T, \xi, \tau) = 0 \tag{1.1.14}$$

for $0 \leq x \leq L$.

  Exercise 1.1.1
 (a) Verify that

$$\gamma(x, t, \xi, \tau) = \delta(x - \xi - c(t - \tau))H(\tau - t) \tag{1.1.15}$$

for $0 \leq x < L$, $0 \leq t \leq T$, where H is the Heaviside unit step function.
 (b) Show that

$$u(\xi, \tau) = u_F(\xi, \tau) \equiv \int_0^T dt \int_0^L dx \, \gamma(x, t, \xi, \tau) F(x, t)$$

$$+ \int_0^L dx \, \gamma(x, 0, \xi, \tau) I(x) + c \int_0^T dt \, \gamma(0, t, \xi, \tau) B(t). \tag{1.1.16}$$

$$\Box$$

Relabeling (1.1.16) yields

$$u_F(x, t) = \int_0^T d\tau \int_0^L d\xi \gamma(\xi, \tau, x, t) F(\xi, \tau)$$

$$+ \int_0^L d\xi \gamma(\xi, 0, x, t) I(\xi) + c \int_0^T d\tau \gamma(0, \tau, x, t) B(\tau). \tag{1.1.17}$$

which is an explicit solution for the "forward model". It is also the prior estimate or "first-guess" or "background" for u.

Note 1. By inspection, u_F depends continuously upon changes to F, I and B; if these change by $O(\epsilon)$, so does u_F.

Note 2. We actually require $I(0) = B(0)$, or else u_F is discontinuous across the phase line $x = ct$, for all t.

We conclude that the forward model (1.1.1)–(1.1.3) is well-posed. Any additional information would overdetermine the system, and a smooth solution would not exist.

Exercise 1.1.2
Code the finite-difference equation

$$u_n^{k+1} = u_n^k - c(\Delta t/\Delta x)\left(u_n^k - u_{n-1}^k\right) + \Delta t\, F_n^k, \tag{1.1.18}$$

where $u_n^k = u(n\Delta x, k\Delta t)$, etc. Perform some numerical integrations. Derive and verify experimentally the Courant–Friedrichs–Lewy stability criterion (Haltiner and Williams, 1980).   □

Exercise 1.1.3
Slow, one-dimensional viscous flow $u = u(x, t)$ is approximately governed by

$$\frac{\partial u}{\partial t} = v\frac{\partial^2 u}{\partial x^2} - \rho^{-1}\frac{\partial p}{\partial x}, \tag{1.1.19}$$

where v is the uniform kinematic viscosity, ρ is the uniform density and $p = p(x, t)$ is the externally imposed pressure gradient. Consider an infinite domain: $-\infty < x < \infty$, and a finite time interval: $0 < t < T$. A suitable initial condition is

$$u(x, 0) = I(x). \tag{1.1.20}$$

Assume that both $\partial p/\partial x$ and I vanish as $|x| \to \infty$.

(a) Derive the following energy integral when both $\partial p/\partial x$ and I vanish everywhere:

$$\frac{d}{dt}\frac{1}{2}\int_{-\infty}^{\infty} u^2\, dx = -v\int_{-\infty}^{\infty}\left(\frac{\partial u}{\partial x}\right)^2 dx. \tag{1.1.21}$$

Hence prove that there is at most one solution for each choice of p and I.

(b) Show that the solution of (1.1.19), (1.1.20) is

$$u(x, t) = -\rho^{-1}\int_0^T d\tau \int_{-\infty}^{\infty} d\xi\, \phi(\xi, \tau, x, t)\frac{\partial p}{\partial x}(\xi, \tau)$$

$$+ \int_{-\infty}^{\infty} d\xi\, \phi(\xi, 0, x, t)I(\xi), \tag{1.1.22}$$

where the Green's function or fundamental solution $\phi(x, t, \xi, \tau)$ satisfies

$$-\frac{\partial \phi}{\partial t} = \nu \frac{\partial^2 \phi}{\partial x^2} + \delta(x - \xi)\delta(t - \tau), \qquad (1.1.23)$$

subject to

$$\phi(x, T, \xi, \tau) = 0 \qquad (1.1.24)$$

for $-\infty < x < \infty$, and

$$\phi(x, t, \xi, \tau) \to 0 \qquad (1.1.25)$$

as $|x| \to \infty$.

(c) Verify that

$$\phi(x, t, \xi, \tau) = \frac{H(\tau - t)e^{\frac{-(x-\xi)^2}{2\nu(\tau-t)}}}{\sqrt{2\pi \nu(\tau - t)}}. \qquad (1.1.26)$$

Notice that the effective range of integration with respect to time in (1.1.22) is $0 < \tau < t$. $\qquad \square$

Exercise 1.1.4

(1) Is quantum mechanics captured mathematically as well-posed problems? See, for example, Schiff (1949, p. 48).

(2) Can well-posed problems have chaotic solutions? $\qquad \square$

1.2   Inverse models

1.2.1   Overdetermined problems

We shall spoil the well-posedness of the forward model examined in §1.1, by introducing additional information about the toy "ocean circulation" field $u(x, t)$. This information will consist of imperfect observations of u at isolated points in space and time, for the sake of simplicity. The forward model becomes overdetermined; it cannot be solved with smooth functions and must be regarded as ill-posed. We shall resolve the ill-posed problem by constructing a weighted, least-squares best-fit to all the information. It will be shown that this best-fit or "generalized inverse" of the ill-posed problem obeys the Euler–Lagrange equations.

1.2.2   Toy ocean data

Let us assume that a finite number M of measurements (observations, data, ...) of u were collected in the bounded "ocean basin" $0 \le x \le L$, during the "cruise" $0 \le t \le T$. The data were collected at the points (x_m, t_m), where $1 \le m \le M$: see Fig. 1.2.1. The

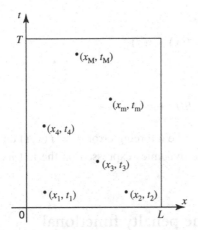

Figure 1.2.1 Toy ocean data.

data are related to the "true" ocean circulation field $u(x, t)$ by

$$d_m = u(x_m, t_m) + \epsilon_m, \qquad (1.2.1)$$

$1 \le m \le M$, where d_m is the datum or recorded value, and $u(x_m, t_m)$ is the true value of the circulation. The measurement error ϵ_m may arise from an imperfect measuring system, or else from mistakenly identifying streamfunction and pressure, for example. On the other hand, if our ocean model were quasigeostrophic and the data included internal waves, then there would also be cause to admit errors in the dynamics.

1.2.3   Failure of the forward solution

Let us now consider how these data relate to the forward problem. If $u_F = u_F(x, t)$ is the forward solution:

$$\frac{\partial u_F}{\partial t} + c \frac{\partial u_F}{\partial x} = F \qquad (1.2.2)$$

for $0 \le x \le L$ and $0 \le t \le T$, with

$$u_F(x, 0) = I(x) \qquad (1.2.3)$$

for $0 \le x \le L$, and

$$u_F(0, t) = B(t) \qquad (1.2.4)$$

for $0 \le t \le T$, then we may expect that

$$u_F(x_m, t_m) \ne d_m \qquad (1.2.5)$$

for at least some m: $1 \le m \le M$. We therefore assume that there are errors in our prior estimates for F, I and B. So the true circulation u must satisfy

$$\frac{\partial u}{\partial t} + c \frac{\partial u}{\partial x} = F + f \qquad (1.2.6)$$

for $0 \le x \le L$ and $0 \le t \le T$,

$$u(x, 0) = I(x) + i(x) \qquad (1.2.7)$$

for $0 \le x \le L$ and

$$u(0, t) = B(t) + b(t) \qquad (1.2.8)$$

for $0 \le t \le T$. Note that what is implied to be a forcing error $f = f(x, t)$ on the rhs of (1.2.6) may actually be an error in the dynamics expressed on the left-hand side (lhs) of (1.2.6).

1.2.4   Least-squares fitting: the penalty functional

We have established that for any choice of $F + f$, $I + i$ and $B + b$, there is a unique solution for u. However, we have only the M data values d_m to guide us and so the error fields f, i and b are undetermined, while the data errors ϵ_m are unknown. We shall seek the field $\hat{u} = \hat{u}(x, t)$ that corresponds to the smallest values for f, i, b and ϵ_m in a weighted, least-squares sense. Specifically, we shall seek the minimum of the quadratic *penalty functional* or *cost functional* $\mathcal{J}$:

$$\mathcal{J} = \mathcal{J}[u] \equiv W_f \int_0^T dt \int_0^L dx \, f(x, t)^2 + W_i \int_0^L dx \, i(x)^2 + W_b \int_0^T dt \, b(t)^2 + w \sum_{m=1}^M \epsilon_m^2,$$
$$(1.2.9)$$

where W_f, W_i, W_b and w are positive weights that we are free to choose. There are more general quadratic forms, but (1.2.9) will suffice for now. The lhs of (1.2.9) expresses the dependence of $\mathcal{J}$ upon u, while the rhs only involves f, i, b and ϵ_m. It is to be understood that the latter are the values that would be obtained, were u substituted into (1.2.1) and (1.2.6)–(1.2.8). These *definitions* could be appended to $\mathcal{J}$ using Lagrange multipliers, but it is simpler just to remember them ourselves. Finally, note that while u is a *field* of values for $0 \le x \le L$ and $0 \le t \le T$, the penalty functional $\mathcal{J}[u]$ is a *single number* for each choice of the *entire field u*.

Rewriting (1.2.9), with f, i, b and ϵ_m replaced by their definitions, yields

$$\mathcal{J}[u] = W_f \int_0^T dt \int_0^L dx \left\{ \frac{\partial u}{\partial t} + c \frac{\partial u}{\partial x} - F \right\}^2 + W_i \int_0^L dx \, \{u(x, 0) - I(x)\}^2$$

$$+ W_b \int_0^T dt \, \{u(0, t) - B(t)\}^2 + w \sum_{m=1}^M \{u(x_m, t_m) - d_m\}^2 . \qquad (1.2.10)$$

The dependence upon u (and upon F, I, B and d_m) is now explicit.

1.2.5   The calculus of variations: the Euler–Lagrange equations

We shall use the calculus of variations (Courant and Hilbert, 1953; Lanczos, 1966) to find a *local extremum* of $\mathcal{J}$. Since $\mathcal{J}$ is quadratic in u and clearly nonnegative, the local extremum must be the global minimum. To begin, let $\hat{u} = \hat{u}(x, t)$ be the local extremum. That is,

$$\mathcal{J}[\hat{u} + \delta u] = \mathcal{J}[\hat{u}] + O(\delta u)^2 \tag{1.2.11}$$

for some small change $\delta u = \delta u(x, t)$. This statement can be made more precise but we shall proceed informally:

$$\delta \mathcal{J} \equiv \mathcal{J}[\hat{u} + \delta u] - \mathcal{J}[\hat{u}]$$

$$= 2W_f \int_0^T dt \int_0^L dx \left\{ \frac{\partial \hat{u}}{\partial t} + c\frac{\partial \hat{u}}{\partial x} - F \right\} \left\{ \frac{\partial \delta u}{\partial t} + c\frac{\partial \delta u}{\partial x} \right\}$$

$$+ 2W_i \int_0^L dx \left\{ \hat{u}(x, 0) - I(x) \right\} \delta u(x, 0) + 2W_b \int_0^T dt \left\{ \hat{u}(0, t) - B(t) \right\} \delta u(0, t)$$

$$+ 2w \sum_{m=1}^M \left\{ \hat{u}(x_m, t_m) - d_m \right\} \delta u(x_m, t_m) + O(\delta u)^2 . \tag{1.2.12}$$

Note 1. F, I, B and d_m have not been allowed to vary; only $\hat{u}$ has been varied.

Note 2. We have assumed that

$$\delta \frac{\partial u}{\partial t}(x, t) = \frac{\partial \delta u}{\partial t}(x, t), \quad \text{etc.} \tag{1.2.13}$$

The lhs of (1.2.13) is a variation of $(\partial u / \partial t)$; the rhs is the time derivative of the variation of u.

For convenience let us introduce the field $\lambda = \lambda(x, t)$:

$$\lambda \equiv W_f \left(\frac{\partial \hat{u}}{\partial t} + c\frac{\partial \hat{u}}{\partial x} - F \right). \tag{1.2.14}$$

Then the first term in $\delta \mathcal{J}$ is

$$2 \int_0^T dt \int_0^L dx \, \lambda \left\{ \frac{\partial \delta u}{\partial t} + c\frac{\partial \delta u}{\partial x} \right\}$$

$$= 2 \left[\int_0^L dx \, \lambda \delta u \right]_{t=0}^{t=T} + 2 \left[\int_0^T dt \, \lambda c \delta u \right]_{x=0}^{x=L}$$

$$+ 2 \int_0^T dt \int_0^L dx \left\{ -\frac{\partial \lambda}{\partial t} - c\frac{\partial \lambda}{\partial x} \right\} \delta u(x, t). \tag{1.2.15}$$

Notice that the last explicit term in $\delta \mathcal{J}$ may be written as

$$
2 \int_0^T dt \int_0^L dx \, w \sum_{m=1}^M \{\hat{u}(x_m, t_m) - d_m\} \delta u(x, t) \delta(x - x_m) \delta(t - t_m), \quad (1.2.16)
$$

where the second and third δs denote Dirac delta functions. We have now expressed $\delta \mathcal{J}$ entirely in terms of $\delta u(x, t)$. None of δu_t, δu_x and $\delta u(x_m, t_m)$ still appear.

We now argue that

$$
\delta \mathcal{J} = O(\delta u)^2, \quad (1.2.17)
$$

implying that $\hat{u}$ is an extremum of $\mathcal{J}$, provided that the coefficients of $\delta u(x, t)$, $\delta u(L, t)$, $\delta u(0, t)$, $\delta u(0, x)$ and $\delta u(T, x)$ all vanish. Examination of (1.2.12), (1.2.15) and (1.2.16) shows that these conditions are, respectively,

$$
-\frac{\partial \lambda}{\partial t} - c \frac{\partial \lambda}{\partial x} + w \sum_{m=1}^M \{\hat{u}_m - d_m\} \delta(x - x_m) \delta(t - t_m) = 0, \quad (1.2.18)
$$

$$
\lambda(L, t) = 0, \quad (1.2.19)
$$

$$
-c\lambda(0, t) + W_b\{\hat{u}(0, t) - B(t)\} = 0, \quad (1.2.20)
$$

$$
-\lambda(x, 0) + W_i\{\hat{u}(x, 0) - I(x)\} = 0, \quad (1.2.21)
$$

$$
\lambda(x, T) = 0, \quad (1.2.22)
$$

where $\hat{u}_m \equiv \hat{u}(x_m, t_m)$. Recall the definition of λ:

$$
\lambda \equiv W_f \left\{ \frac{\partial \hat{u}}{\partial t} + c \frac{\partial \hat{u}}{\partial x} - F \right\}. \quad (1.2.23)
$$

These conditions (1.2.18)–(1.2.23) constitute the Euler–Lagrange equations for local extrema of the penalty functional $\mathcal{J}$ defined in (1.2.10). How shall we untangle them, to find our best-fit estimate $\hat{u}$ of the ocean circulation u?

Note 1. Substituting (1.2.23) into (1.2.18) yields "the" Euler–Lagrange equation familiar to physicists.

Note 2. Students sometimes derive (1.2.12) from (1.2.10) by expanding the squares in the integrand, evaluating at $\hat{u} + \delta u$ and at $\hat{u}$, and then subtracting. It is less tedious to calculate as follows:

$$
\delta \mathcal{J}[u] \Big|_{u=\hat{u}} = \delta W_f \int_0^T dt \int_0^L dx \left\{ \frac{\partial u}{\partial t} + c \frac{\partial u}{\partial x} - F \right\}^2 + \cdots
$$

$$
= W_f \int_0^T dt \int_0^L dx \, \delta \left(\left\{ \frac{\partial u}{\partial t} + c \frac{\partial u}{\partial x} - F \right\}^2 \right) + \cdots
$$

$$= W_f \int\limits_0^T dt \int\limits_0^L dx\, 2 \left\{ \frac{\partial \hat{u}}{\partial t} + c \frac{\partial \hat{u}}{\partial x} - F \right\} \delta \left\{ \frac{\partial u}{\partial t} + c \frac{\partial u}{\partial x} - F \right\} + \cdots$$

$$= 2 W_f \int\limits_0^T dt \int\limits_0^L dx \left\{ \frac{\partial \hat{u}}{\partial t} + c \frac{\partial \hat{u}}{\partial x} - F \right\} \left\{ \frac{\partial \delta u}{\partial t} + c \frac{\partial \delta u}{\partial x} \right\} + \cdots,$$

$$(1.2.24)$$

as in (1.2.12).

Exercise 1.2.1 (requires care)
Consider the integral

$$\mathcal{I} = \int\limits_0^T dt \int\limits_0^L dx\, \lambda^2 . \qquad (1.2.25)$$

Substitute for one of the factors of λ in (1.2.25), using (1.2.23). Integrate by parts, and use (1.2.18)–(1.2.22). Conclude that if W_f, W_b, W_i and $w > 0$, then the Euler–Lagrange equations (1.2.18)–(1.2.23) have a unique solution. Discuss the case $W_i = 0$; it occurs widely in the published literature (Bennett and Miller, 1991).   ☐

Exercise 1.2.2
Consider slow, viscous flow driven by an externally imposed pressure gradient, as in Exercise 1.1.3. Assume measurements of u are available, as §1.2.2. Resolve this ill-posed problem by defining a generalized inverse in terms of a weighted, least-squares best fit to all the information. Derive the Euler–Lagrange equations, and prove that they have at most one solution.   ☐

Exercise 1.2.3
Introduce a forcing error $\Delta t f_n^k$ into the finite-difference model (1.1.18). By analogy to (1.2.9), a simple penalty function is

$$J[u] = W_f \sum_k \sum_n \left(f_n^k \right)^2 \Delta x \Delta t + \cdots, \qquad (1.2.26)$$

where the ellipsis indicates initial penalties, etc., that will be considered below in stages, as will the ranges of the summations in (1.2.26).

(i) Show that the Euler–Lagrange equation for extrema of J with respect to variations of u_n^k is

$$\lambda_n^{k-1} - \lambda_n^k - c(\Delta t / \Delta x)(\lambda_n^k - \lambda_{n+1}^k) = \cdots, \qquad (1.2.27)$$

where the ellipses indicate contributions from variations of data penalties in (1.2.26).

(ii) The range of summation over the time index k in (1.2.26) is $0 \leq k \leq K - 1$, where $K \, \Delta t = T$. By analogy to (1.2.9), a simple initial penalty is

$$J[u] = \cdots + W_i \sum_n \left(u_n^0 - I_n \right)^2 + \cdots. \qquad (1.2.28)$$

Show that, for extrema with respect to u_n^0 and u_n^K,

$$-\lambda_n^0 + W_i \left\{ \hat{u}_n^0 - I_n \right\} = 0 \qquad (1.2.29)$$

and

$$\lambda_n^{K-1} = 0, \qquad (1.2.30)$$

respectively. Compare these with (1.2.21) and (1.2.22).
(iii) Choose a range of summation over the space index n in (1.2.26), and prescribe a simple boundary penalty analogous to that in (1.2.9). Derive extremal conditions analogous to (1.2.19) and (1.2.20).
(iv) Assume that there are M measurements of u_n^k, that is, measured values of $u(x, t)$ at grid points in space and time. Prescribe a simple data penalty as in (1.2.9) and derive the contributions to (1.2.27) from variations of this data penalty.
Hint: Replace the Dirac delta functions $\delta(x_n - x_m)$ and $\delta(t_k - t_j)$ with $(\Delta x)^{-1} \delta_{nm}$ and $(\Delta t)^{-1} \delta_{kj}$ respectively, where δ_{nm} is the Kronecker delta:

$$\delta_{nm} = \begin{cases} 1 & n = m \\ 0 & n \neq m. \end{cases} \qquad (1.2.31)$$

$\square$

1.3 Solving the Euler–Lagrange equations using representers

1.3.1 Least-squares fitting by explicit solution of extremal conditions

The mixed initial-boundary value problem (1.1.1)–(1.1.3) for the first-order wave equation, together with the data (1.2.1) and the simple least-squares penalty functional (1.2.10), have led us to the awkward system of Euler–Lagrange equations (1.2.18)–(1.2.23). The solution is the best-fit ocean circulation $\hat{u}$. It may be obtained explicitly, by an intricate construction involving "representer functions". The effort is rewarded not only with structural insight, but also with enormous gains in computational efficiency compared to conventional minimization of (1.2.10) using gradient information.

1.3.2   The Euler–Lagrange equations are a two-point boundary value problem in time

After a little reordering, the Euler–Lagrange equations for local extrema $\hat{u}$ of the penalty functional $\mathcal{J}[u]$ are:

$$(B)\begin{cases} -\dfrac{\partial \lambda}{\partial t} - c\dfrac{\partial \lambda}{\partial x} = -w\displaystyle\sum_{m=1}^{M}\{\hat{u}_m - d_m\}\delta(x - x_m)\delta(t - t_m) & (1.3.1) \\[2mm] \lambda(x, T) = 0 & (1.3.2) \\[1mm] \lambda(L, t) = 0, & (1.3.3) \end{cases}$$

$$(F)\begin{cases} \dfrac{\partial \hat{u}}{\partial t} + c\dfrac{\partial \hat{u}}{\partial x} = F + W_f^{-1}\lambda & (1.3.4) \\[2mm] \hat{u}(x, 0) = I(x) + W_i^{-1}\lambda(x, 0) & (1.3.5) \\[2mm] \hat{u}(0, t) = B(t) + cW_b^{-1}\lambda(0, t). & (1.3.6) \end{cases}$$

Note 1. Our best estimates for f, i and b are

$$\hat{f}(x, t) \equiv W_f^{-1}\lambda(x, t), \quad \hat{i}(x) \equiv W_i^{-1}\lambda(x, 0), \quad \hat{b}(t) \equiv cW_b^{-1}\lambda(0, t).$$
$$(1.3.7)$$

Note 2. Eq. (1.3.1) is known as the "backward" or "adjoint" equation.

Note 3. At first glance, it would seem that we could proceed by integrating the system (B) "backwards in time and to the left" (see Fig. 1.2.1), yielding $\hat{\lambda}(x, t)$, $\hat{\lambda}(0, t)$ and $\hat{\lambda}(x, 0)$. Then we could integrate the system (F) "forwards and to the right" (see Fig. 1.2.1), yielding the ocean circulation estimate $\hat{u} = \hat{u}(x, t)$. However, after reexamining (1.3.1), we see that it is necessary to know $\hat{u}(x_m, t_m)$ in order to integrate (B). *The Euler–Lagrange equations do not consist of two initial-value problems; they constitute a single, two-point boundary value problem in the time interval* $0 \leq t \leq T$.

1.3.3   Representer functions: the explicit solution and the reproducing kernel

Let us introduce the representer functions. There are M of them, denoted by $r_m(x, t)$, $1 \leq m \leq M$. Each has an "adjoint" $\alpha_m(x, t)$, satisfying

$$(B_m)\begin{cases} -\dfrac{\partial \alpha_m}{\partial t} - c\dfrac{\partial \alpha_m}{\partial x} = \delta(x - x_m)\delta(t - t_m) & (1.3.8) \\[2mm] \alpha_m(x, T) = 0 & (1.3.9) \\[1mm] \alpha_m(L, t) = 0. & (1.3.10) \end{cases}$$

As a consequence of the single impulse on the rhs of (1.3.8) being "bare", we may integrate (B_m) "backwards and to the left", yielding $\alpha_m(x, t)$. We may then solve for

r_m by integrating (F_m) "forward and to the right":

$$(F_m) \begin{cases} \dfrac{\partial r_m}{\partial t} + c\dfrac{\partial r_m}{\partial x} = W_f^{-1}\alpha_m & (1.3.11) \\[2mm] r_m(x,0) = W_i^{-1}\alpha_m(x,0) & (1.3.12) \\[2mm] r_m(0,t) = cW_b^{-1}\alpha_m(0,t). & (1.3.13) \end{cases}$$

Next, we seek a solution of (1.3.1)–(1.3.6) in the form

$$\hat{u}(x,t) = u_F(x,t) + \sum_{m=1}^{M} \beta_m r_m(x,t), \tag{1.3.14}$$

where u_F is the prior estimate (the solution of the forward model (1.2.2)–(1.2.4)), and the β_m are unknown constants. If we substitute (1.3.14) into (1.3.4), and derive

$$D\hat{u} = Du_F + \sum_{m=1}^{M} \beta_m Dr_m \tag{1.3.15}$$

$$= F + W_f^{-1}\sum_{m=1}^{M} \beta_m \alpha_m, \tag{1.3.16}$$

where $D = \frac{\partial}{\partial t} + c\frac{\partial}{\partial x}$, we find that

$$\lambda \equiv W_f\{D\hat{u} - F\} = \sum_{m=1}^{M} \beta_m \alpha_m. \tag{1.3.17}$$

Furthermore,

$$-D\lambda = -\sum_{m=1}^{M} \beta_m D\alpha_m$$

$$= \sum_{m=1}^{M} \beta_m \delta(x - x_m)\delta(t - t_m) \tag{1.3.18}$$

$$= -w\sum_{m=1}^{M} \{\hat{u}_m - d_m\}\delta(x - x_m)\delta(t - t_m), \tag{1.3.19}$$

by virtue of (1.3.1). Equating coefficients of the impulses, we obtain the optimal choices $\hat{\beta}_m$ for the representer coefficients β_m:

$$\beta_m = \hat{\beta}_m \equiv -w\{\hat{u}_m - d_m\} \tag{1.3.20}$$

for $1 \leq m \leq M$. Substituting again for $\hat{u}_m$ yields

$$\hat{\beta}_m = -w\left\{ u_{F_m} + \sum_{l=1}^{M} \hat{\beta}_l r_{lm} - d_m \right\}, \tag{1.3.21}$$

where $u_{F_m} \equiv u_F(x_m, t_m)$ and $r_{lm} \equiv r_l(x_m, t_m)$.

Hence

$$\sum_{l=1}^{M}(r_{lm} + w^{-1}\delta_{lm})\hat{\beta}_l = h_m \equiv d_m - u_{F_m}, \qquad (1.3.22)$$

where δ_{lm} is the Kronecker delta. In matrix notation, the M equations (1.3.22) for the M representer coefficients $\hat{\beta}_m$ become

$$(\mathbf{R} + w^{-1}\mathbf{I})\hat{\beta} = \mathbf{h} \equiv \mathbf{d} - \mathbf{u}_F. \qquad (1.3.23)$$

Note 1. The rhs $\mathbf{h}$ is known; it is the data vector minus the vector of measured values of the prior estimate.

Note 2. The diagonal weight matrix $w\mathbf{I}$ is readily generalized to symmetric positive definite matrices $\mathbf{w}$.

Note 3. The lth column of the $M \times M$ "representer matrix" $\mathbf{R}$ consists of the M measured values of the lth representer function $r_l(x, t)$.

Note 4. It will be shown (see (1.3.32)) that $\mathbf{R}$ is symmetric: $\mathbf{R} = \mathbf{R}^{\text{T}}$.

Note 5. The generalized inverse problem of finding the field $\hat{u} = \hat{u}(x, t)$, where $0 \leq x \leq L$ and $0 \leq t \leq T$, has been exactly reduced to the problem of inverting an $M \times M$ matrix, in order to find the M representer coefficients $\hat{\beta}$.

Finally, we have an explicit solution for $\hat{u}$:

$$\hat{u}(x, t) = u_F(x, t) + (\mathbf{d} - \mathbf{u}_F)^{\text{T}}(\mathbf{R} + \mathbf{w}^{-1})^{-1}\mathbf{r}(x, t). \qquad (1.3.24)$$

It was established in §1.1 that the forward model (1.1.1)–(1.1.3) has a unique solution for each choice of the inputs. Accordingly, the partial differential operator in (1.1.1), the initial operator in (1.1.2) and the boundary operator in (1.1.3) constitute a nonsingular operator. It may be inverted; the inverse operator is expressed explicitly in (1.1.17) with the Green's function γ. Introducing the measurement operators as in (1.2.1) yields a problem with no solution, thus the operator comprising those in (1.1.1)–(1.1.3) and (1.2.1) is singular; it is not invertible in the regular sense. However, a generalized inverse has been defined in the weighted least-squares sense of (1.2.10), and is explicitly expressed in (1.3.24) with the representers for the penalty functional (1.2.10), and with the Green's function for the nonsingular operator. Recall that u_F is given by (1.1.17), although it will in practice be computed by numerical integration of (1.2.2)–(1.2.4). In an abuse of language, we shall refer to the best-fit $\hat{u}$ given by (1.3.24) as the generalized inverse estimate, or simply the inverse.

Exercise 1.3.1

Verify that the initial condition (1.3.5) and boundary conditions (1.3.6) are satisfied.

□

In summary, the steps for solving the Euler–Lagrange equations are:

(1) calculate $u_F(x, t)$ and hence $\mathbf{u}_F$;
(2) calculate $\mathbf{r}(x, t)$ and hence $\mathbf{R}$;
(3) invert $\mathbf{P} \equiv \mathbf{R} + \mathbf{w}^{-1}$;
(4) assemble (1.3.24).

Note 1. $u_F(x, t)$ depends upon the "dynamics", the initial operator, the boundary
operator and the choices for F, I and B.

Note 2. $\mathbf{u}_F$ depends upon u_F and the "observing network" $\{(x_m, t_m)\}_{m=1}^{M}$.

Note 3. $\mathbf{r}$ depends upon the dynamics, the initial operator, the boundary operator, the
observing network and the inverted weights W_f^{-1}, W_i^{-1}, W_b^{-1}.

Note 4. $\hat{\beta}$ depends upon $\mathbf{R}$, the inverse of the data weight $\mathbf{w}$, and the prior data misfit
$\mathbf{h} \equiv \mathbf{d} - \mathbf{u}_F$.

Note 5. See Fig. 3.1.1 for a "time chart" implementing the representer solution.

Exercise 1.3.2
Express λ, $\hat{f}$, $\hat{\imath}$ and $\hat{b}$ using representer functions and their adjoints.   □

Exercise 1.3.3 (trivial)
Show that

$$\mathcal{J}_F \equiv \mathcal{J}[u_F] = \mathbf{h}^\mathsf{T} \mathbf{w} \mathbf{h}. \tag{1.3.25}$$

□

Exercise 1.3.4 (nontrivial)
Show that

(i) $\hat{\mathcal{J}} \equiv \mathcal{J}[\hat{u}] = \mathbf{h}^\mathsf{T} \mathbf{P}^{-1} \mathbf{h},$ (1.3.26)

(ii) $\hat{\mathcal{J}}_{\text{data}} \equiv (\mathbf{d} - \hat{\mathbf{u}})^\mathsf{T} \mathbf{w} (\mathbf{d} - \hat{\mathbf{u}}) = \mathbf{h}^\mathsf{T} \mathbf{P}^{-1} \mathbf{w}^{-1} \mathbf{P}^{-1} \mathbf{h},$ (1.3.27)

and

(iii) $\hat{\mathcal{J}}_{\text{mod}} \equiv \hat{\mathcal{J}} - \hat{\mathcal{J}}_{\text{data}} = \mathbf{h}^\mathsf{T} \mathbf{P}^{-1} \mathbf{R} \mathbf{P}^{-1} \mathbf{h}.$ (1.3.28)

Note that $\hat{\mathcal{J}}_{\text{mod}}$ is the sum of dynamical, initial and boundary penalties.   □

Let us now prove that the representer matrix is symmetric: $\mathbf{R} = \mathbf{R}^\mathsf{T}$.

First, recall that the adjoint representer $\alpha_m(x, t)$ for a point measurement at (x_m, t_m)
is just the Green's function $\gamma(x, t, x_m, t_m)$, where $\gamma(x, t, y, s)$ satisfies

$$-\frac{\partial \gamma}{\partial t} - c \frac{\partial \gamma}{\partial x} = \delta(x - y)\delta(t - s), \tag{1.3.29}$$

subject to $\gamma = 0$ at $t = T$, $\gamma = 0$ at $x = L$. Now let $\Gamma(x, t, y, s)$ satisfy

$$\frac{\partial}{\partial t}\Gamma + c\frac{\partial}{\partial x}\Gamma = W_f^{-1}\gamma, \qquad (1.3.30)$$

subject to $\Gamma = W_i^{-1}\gamma$ at $t = 0$, and $\Gamma = cW_b^{-1}\gamma$ at $x = 0$. Thus $r_m(x, t) = \Gamma(x, t, x_m, t_m)$.

Exercise 1.3.5 (Bennett, 1992)
Show that

$$\Gamma(x, t, y, s) = W_f^{-1}\int dz \int dr\, \gamma(z, r, x, t)\gamma(z, r, y, s)$$
$$+ W_i^{-1}\int dz\, \gamma(z, 0, x, t)\gamma(z, 0, y, s)$$
$$+ c^2 W_b^{-1}\int dr\, \gamma(0, r, x, t)\gamma(0, r, y, s). \qquad (1.3.31)$$

Hence representers are not Green's functions; rather they are "squares" of Green's functions. Note that Γ is symmetric, but γ is *not* symmetric.   □

Finally we deduce that

$$r_{lm} \equiv r_l(x_m, t_m) = \Gamma(x_m, t_m, x_l, t_l)$$
$$= \Gamma(x_l, t_l, x_m, t_m)$$
$$= r_m(x_l, t_l)$$
$$\equiv r_{ml}. \qquad (1.3.32)$$

That is, $\mathbf{R} = \mathbf{R}^T$. Note that Γ is known as a "reproducing kernel" or "rk", for reasons given in §2.1.

1.4   Some limiting choices of weights: "weak" and "strong" constraints

1.4.1   Diagonal data weight matrices, for simplicity

The parade of formulae in the previous sections should become more meaningful as we explore some limiting choices for the weights. We shall assume that the data weight matrix is diagonal:

$$\mathbf{w} = w\mathbf{I}, \qquad (1.4.1)$$

in order to avoid technicalities such as the norm of a matrix. Note that (1.4.1) implies

$$\mathbf{w}^{-1} = w^{-1}\mathbf{I}. \qquad (1.4.2)$$

1.4.2   Perfect data

If we believe that the data are perfectly accurate, then we should give infinite weight to them. In this case we hope that the inverse estimates agree exactly with the data, at the measurement sites. Let us therefore consider the limit: $w \rightarrow \infty$.

Hence

$$\mathbf{P} \equiv \mathbf{R} + w^{-1}\mathbf{I} \rightarrow \mathbf{R}, \tag{1.4.3}$$

$$\hat{\beta} \rightarrow \mathbf{R}^{-1}\mathbf{h}, \tag{1.4.4}$$

and

$$\hat{u}(x, t) \rightarrow u_F(x, t) + \mathbf{r}(x, t)^{\mathsf{T}}\mathbf{R}^{-1}\mathbf{h}. \tag{1.4.5}$$

Measuring both sides of (1.4.5) yields

$$\hat{\mathbf{u}} \rightarrow \mathbf{u}_F + \mathbf{R}^{\mathsf{T}}\mathbf{R}^{-1}\mathbf{h}, \tag{1.4.6}$$

$$= \mathbf{u}_F + \mathbf{h} \tag{1.4.7}$$

$$= \mathbf{u}_F + (\mathbf{d} - \mathbf{u}_F) \tag{1.4.8}$$

$$= \mathbf{d}, \tag{1.4.9}$$

as required. Note that we have used the symmetry of the representer matrix: $\mathbf{R}^{\mathsf{T}} = \mathbf{R}$. In this limit, the inverse estimate *interpolates* the data.

1.4.3   Worthless data

Now suppose that we believe the data are worthless, that is, we have no information about the magnitude of the data errors. In practice we always have some idea: the errors in altimetry data do not exceed the height of the orbit of the satellite, but that is infinite by any hydrographic standard. We should therefore consider the limit: $w \rightarrow 0$. Then

$$\mathbf{P}^{-1} = (\mathbf{R} + w^{-1}\mathbf{I})^{-1} \rightarrow \mathbf{0}, \tag{1.4.10}$$

hence

$$\hat{u}(x, t) \rightarrow u_F(x, t). \tag{1.4.11}$$

That is, the data have no influence on the inverse estimate, as would be desirable.

1.4.4   Rescaling the penalty functional

The Euler–Lagrange equations for local extrema of $\mathcal{J}[u]$ are also those for local extrema of $2\mathcal{J}[u]$. This is true even if the dynamics and observing systems are nonlinear, or if $\mathcal{J}$ is not quadratic. Thus the limiting cases: $w \rightarrow \infty$, $w \rightarrow 0$ really refer to w/W_f, w/W_i, w/W_b all $\rightarrow \infty$, or all $\rightarrow 0$. That is, they refer to the relative weighting of the various information. However, the prior and posterior functional values $\mathcal{J}_F \equiv \mathcal{J}[u_F]$, and $\hat{\mathcal{J}} \equiv \mathcal{J}[\hat{u}]$ do depend upon the absolute values of the weights. This will be crucial later, when we interpret these numbers as test statistics for the model as a formal hypothesis about the ocean.

1.4.5   Perfect dynamics: Lagrange multipliers for strong constraints

The admission of the error field $f = f(x, t)$ in (1.2.6), and the inclusion of $W_f \int_0^L dx \int_0^T dt \, f^2$ in the penalty functional (1.2.9), leads to the model being described as a "weak constraint" upon the inversion process (Sasaki, 1970). The model may alternatively be imposed as a "strong constraint". In that case, the penalty functional is

$$\mathcal{K}[u] = W_i \int_0^L dx \, i(x)^2 + W_b \int dt \, b(t)^2 + w \sum_{m=1}^M \epsilon_m^2. \qquad (1.4.12)$$

Compare (1.4.12) and (1.2.9): i, b and ϵ_m are defined as before by (1.2.7), (1.2.8) and (1.2.1) respectively, but now we require that $u = u(x, t)$ satisfy (1.1.1) exactly. This requirement may be met in the search for the minimum of $\mathcal{K}$, by appending the strong constraint (1.1.1) to $\mathcal{K}$ using a Lagrange multiplier field $\psi = \psi(x, t)$:

$$\mathcal{L}[u, \psi] = \mathcal{K}[u] + 2 \int_0^L dx \int_0^T dt \, \psi(x, t) \left\{ \frac{\partial u}{\partial t}(x, t) + c \frac{\partial u}{\partial x}(x, t) - F(x, t) \right\}. \qquad (1.4.13)$$

The factor of two will be seen to be convenient. Note that the augmented penalty functional $\mathcal{L}$ depends on u and ψ, which may vary independently. The total variation in $\mathcal{L}$ is

$$\delta \mathcal{L} = \delta \mathcal{K} + 2 \int_0^L dx \int_0^T dt \, \delta \psi \left(\frac{\partial u}{\partial t} + c \frac{\partial u}{\partial x} - F \right)$$

$$+ 2 \int_0^L dx \int_0^T dt \, \psi \left(\delta \frac{\partial u}{\partial t} + c \delta \frac{\partial u}{\partial x} \right) + O(\delta^2). \qquad (1.4.14)$$

If the pair of fields $\hat{\psi} = \hat{\psi}(x, t)$ and $\hat{u} = \hat{u}(x, t)$ extremize $\mathcal{L}$ for arbitrary variations $\delta \psi$ and δu, then

$$-\frac{\partial \psi}{\partial t} - c \frac{\partial \psi}{\partial x} = -w \sum_{m=1}^M \{\hat{u}_m - d_m\} \delta(x - x_m) \delta(t - t_m) \qquad (1.4.15)$$

$$\psi(x, T) = 0 \qquad (1.4.16)$$

$$\psi(L, t) = 0 \qquad (1.4.17)$$

$$\frac{\partial \hat{u}}{\partial t} + c \frac{\partial \hat{u}}{\partial x} = F \qquad (1.4.18)$$

$$\hat{u}(x, 0) = I(x) + W_i^{-1} \psi(x, 0) \qquad (1.4.19)$$

$$\hat{u}(0, t) = B(t) + c W_b^{-1} \psi(0, t). \qquad (1.4.20)$$

The strong constraint (1.4.18) is immediately recovered from (1.4.14), if $\delta\mathcal{L} = 0$ and $\delta\psi$ is arbitrary. The other Euler–Lagrange conditions are recovered as in §1.2. Comparing (1.4.15)–(1.4.20) with (1.3.1)–(1.3.6) establishes that

$$\psi(x, t) = \lim_{W_f \to \infty} \lambda(x, t). \qquad (1.4.21)$$

It would seem from (1.3.1)–(1.3.3) that λ is independent of W_f, but there is an implicit dependence through $\hat{u}_m$, $1 \le m \le M$, in (1.3.1). Thus, we may recover the "strong constraint" inverse from the "weak constraint" inverse in the limit as $W_f \to \infty$.

1.5   Regularity of the inverse estimate

1.5.1   Physical realizability

Thus far our construction has been formal: we paid no attention to the physical realizability of the inverse estimate $\hat{u}$. We shall now see that $\hat{u}$ is in fact unrealistic, unless we make more interesting choices for the weights W_f, W_i and W_b.

1.5.2   Regularity of the Green's functions and the adjoint representer functions

Consider "the" Euler–Lagrange equation:

$$-\frac{\partial\lambda}{\partial t} - c\frac{\partial\lambda}{\partial x} = w\sum_{m=1}^{M}(d_m - \hat{u}_m)\delta(x - x_m)\delta(t - t_m). \qquad (1.5.1)$$

In fact, just consider the equation for an adjoint representer function:

$$-\frac{\partial\alpha_m}{\partial t} - c\frac{\partial\alpha_m}{\partial x} = \delta(x - x_m)\delta(t - t_m), \qquad (1.5.2)$$

subject to

$$\alpha_m(x, T) = 0, \quad \alpha_m(L, t) = 0. \qquad (1.5.3)$$

The solution is the Green's function:

$$\alpha_m(x, t) = \gamma(x, t, x_m, t_m)$$

$$= \delta(x - x_m - c(t - t_m))H(t_m - t), \qquad (1.5.4)$$

and $\lambda(x, t) = \beta^{\mathrm{T}}\alpha(x, t)$. Clearly the α_m and hence λ are singular, and not just at the data points (x_m, t_m): see Fig. 1.5.1.

Now $\hat{u}$ obeys

$$\frac{\partial\hat{u}}{\partial t} + c\frac{\partial\hat{u}}{\partial x} = F + \hat{f} = F + W_f^{-1}\lambda, \qquad (1.5.5)$$

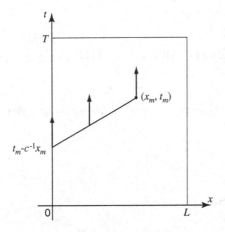

Figure 1.5.1 Support of $\alpha_m(x, t)$. The arrows (the delta functions) are normal to the page.

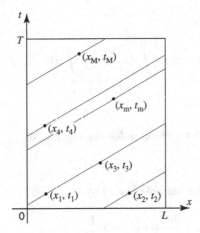

Figure 1.5.2 Support of singularities in $\hat{u}(x, t)$.

subject to the initial and boundary conditions

$$\hat{u} = I + \hat{\imath} = I + W_i^{-1}\lambda, \quad \hat{u} = B + \hat{b} = B + cW_b^{-1}\lambda. \tag{1.5.6}$$

So our estimates of $\hat{f}$, $\hat{\imath}$ and $\hat{b}$ are singular. There is neither dispersion nor diffusion in our "toy" ocean dynamics, so $\hat{u}$ is also singular: see Fig. 1.5.2. This is hardly a satisfactory combination of dynamics and data!

Exercise 1.5.1

Express r_m and $\hat{u}$ using the Green's function γ. □

1.5.3 Nondiagonal weighting: kernel inverses of weights

We want the data to influence the circulation at remote places and times, so we should give weight to products of residuals at remote places and times. We therefore generalize the penalty functional (1.2.9) to

$$\mathcal{J}[u] = \int\limits_0^T dt \int\limits_0^T ds \int\limits_0^L dx \int\limits_0^L dy \, f(x,t) W_f(x,t,y,s) f(y,s)$$

$$+ \int\limits_0^L dx \int\limits_0^L dy \, i(x) W_i(x,y) i(y) + \int\limits_0^T dt \int\limits_0^T ds \, b(t) W_b(t,s) b(s)$$

$$+ \sum_{l=1}^{M} \sum_{m=1}^{M} \epsilon_l w_{lm} \epsilon_m. \tag{1.5.7}$$

Thus our previous, trivial choices were

$$W_f(x,t,y,s) = W_f \cdot \delta(x-y)\delta(t-s), \quad \text{etc.} \tag{1.5.8}$$

The notations

$$\bullet \equiv \int\limits_0^T dt \int\limits_0^L dx, \quad \circ \equiv \int\limits_0^L dx, \quad * \equiv \int\limits_0^T dt$$

allow us to write $\mathcal{J}$ more compactly as

$$\mathcal{J}[u] = f \bullet W_f \bullet f + i \circ W_i \circ i + b * W_b * b + \epsilon^{\mathsf{T}} \mathbf{w} \epsilon. \tag{1.5.9}$$

Exercise 1.5.2

Define the weighted residual or adjoint variable $\lambda(x,t)$ by

$$\lambda \equiv W_f \bullet \left\{ \frac{\partial \hat{u}}{\partial t} + c \frac{\partial \hat{u}}{\partial x} - F \right\}. \tag{1.5.10}$$

Then show that the Euler–Lagrange equations for minima of (1.5.9) are just as before.   □

Exercise 1.5.3

Define C_f, the inverse of W_f, by

$$C_f \bullet W_f \equiv \int\limits_0^T dr \int\limits_0^L dz \, C_f(x,t,z,r) W_f(z,r,y,s) \tag{1.5.11}$$

$$= \delta(x-y)\delta(t-s). \tag{1.5.12}$$

Define C_i and C_b analogously, and define $\mathbf{C}_\epsilon$ by

$$\mathbf{w}\mathbf{C}_\epsilon = \mathbf{I}. \tag{1.5.13}$$

Each entity in (1.5.13) is an $M \times M$ matrix. Now, write out the representer solution of the Euler–Lagrange equations. Verify that the solution only requires C_f, C_i, C_b and C_ϵ; that is, it does not require their inverses, the weights W_f, W_i, W_b and $\mathbf{w}$.   □

1.5.4 The inverse weights smooth the residuals

The inverse estimate $\hat{u}$ obeys

$$\frac{\partial \hat{u}}{\partial t}(x, t) + c\frac{\partial \hat{u}}{\partial x}(x, t) = F(x, t) + (C_f \bullet \lambda)(x, t)$$

$$= F(x, t) + \int_0^T ds \int_0^L dy\, C_f(x, t, y, s)\lambda(y, s), \quad (1.5.14)$$

subject to

$$\hat{u}(x, 0) = I(x) + (C_i \circ \lambda)(x, 0), \quad (1.5.15)$$

and

$$\hat{u}(0, t) = B(t) + c(C_b * \lambda)(0, t). \quad (1.5.16)$$

The supposition is that C_f, C_i and C_b smooth the singular behavior of λ, yielding regular estimates for $\hat{f} \equiv C_f \bullet \lambda$, $\hat{\imath} \equiv C_i \circ \lambda$ and $\hat{b} = cC_b * \lambda$, leading in turn to a regular estimate $\hat{u}$ for the ocean circulation.

In summary, we should avoid "diagonal" weighting.

Note 1. The adjoint variables α and λ remain singular, but $\mathbf{r}$ and u should become regular.

Note 2. Evaluation of the convolutions in (1.5.14), (1.5.15) and (1.5.16) at each position and time is potentially very expensive: consider *three* space dimensions and time.

Note 3. Functional analysis sheds much light on smoothness: see §2.6.

Exercise 1.5.4

Consider slow, viscous flow as discussed in Exercises 1.1.3 and 1.2.2. Construct both the adjoint representers α and the representers $\mathbf{r}$, using the Green's function ϕ given in (1.1.26). How smooth are α and $\mathbf{r}$? Is nondiagonal weighting of either the dynamical penalty or the initial penalty necessary?   □

Exercise 1.5.5

Generalize the definition of the rk Γ given in (1.3.30) *et seq.* Prove that

(i) $r_m(x, t) = \Gamma(x, t, x_m, t_m),$ (1.5.17)
      for $1 \leq m \leq M$;

(ii) $\Gamma = \gamma \bullet C_f \bullet \gamma + \gamma \circ C_i \circ \gamma + c^2 \gamma * C_b * \gamma.$ (1.5.18)

$\square$

Note:

The adjoint equations (1.3.1)–(1.3.3) and forward equations (1.5.14)–(1.5.16),
which constitute the most general form of the Euler–Lagrange equations developed in §1.2 and §1.5, are restated for convenience in §4.2 as (4.2.1)–(4.2.6).

Chapter 2

Interpretation

The calculus of variations uses Green's functions and representers to express the best fit to a linear model and data. Mathematical construction of the representers is devious, and the meaning of the representer solution to the "control problem" of Chapter 1 is not obvious. There is a geometrical interpretation, in terms of observable and unobservable degrees of freedom. Unobservability defines an orthogonality, and the representers span a finite-dimensional subspace of the space of all model solutions or "circulations". The representers are in fact the observable degrees of freedom.

A statistical interpretation is also available: if the unknown errors in the model are regarded as random fields having prescribed means and covariances, then the representers are related, via the measurement processes, to the covariances of the circulations. Thus the representer solution to the variational problem is also the optimal linear interpolation, in time and space, of data from multivariate, inhomogeneous and nonstationary random fields. The minimal value of the penalty functional that defines the generalized inverse or control problem is a random number. It is the χ^2 variable, if the prescribed error means or covariances are correct, and has one degree of freedom per datum. Measurements need not be pointwise values of the circulation; representers along with their geometrical and statistical interpretations may be constructed for all bounded linear measurement functionals.

Analysis of the conditioning of the determination of the representer amplitudes reveals those degrees of freedom which are the most stable with respect to the observations. This characterization also indicates the efficiency of the observing system – the fewer unstable degrees of freedom, the better.

Interpreting the variational formulation is completed by demonstrating the relationship between weights, covariances and roughness penalties.

2.1   Geometrical interpretation

2.1.1   Alternatives to the calculus of variations

After formulating the penalty functional that defines the best fit to our model and our data, we found a local extremum using the theory of the calculus of the first variation. Specifically, we derived the Euler–Lagrange equations, and explicitly expressed their solution with representers. These functions were defined as special and directly calculable solutions of Euler–Lagrange-like equations. We shall now construct the same extremum for the penalty functional using Hilbert Space theory (Yoshida, 1980). This geometrical construction reveals the efficiency of minimization algorithms based on the Euler–Lagrange equations.

Exercise 2.1.1

How do we know that we shall find the same extremum?   ☐

2.1.2   Inner products

We begin by defining an inner product for two "ocean circulations" $u = u(x, t)$ and $v = v(x, t)$:

$$
\langle u, v \rangle \equiv \int_0^T dt \int_0^T ds \int_0^L dx \int_0^L dy \left\{ \left(\frac{\partial}{\partial t} + c \frac{\partial}{\partial x} \right) u(x, t) \right\}
$$

$$
\times W_f(x, t, y, s) \left\{ \left(\frac{\partial}{\partial s} + c \frac{\partial}{\partial y} \right) v(y, s) \right\}
$$

$$
+ \int_0^L dx \int_0^L dy \, u(x, 0) W_i(x, y) v(y, 0)
$$

$$
+ \int_0^T dt \int_0^T ds \, u(0, t) W_b(t, s) v(0, s)
$$

$$
= f_u \bullet W_f \bullet f_v + u \circ W_i \circ v + u * W_b * v, \qquad (2.1.1)
$$

where f_u is the residual for u (Bennett, 1992).

Exercise 2.1.2

Verify that $\langle \, , \, \rangle$ is an inner product, that is:

  (i) $\langle u, v \rangle = \langle v, u \rangle$ (assume the W are symmetric),
  (ii) $\langle cu + dw, v \rangle = c \langle u, v \rangle + d \langle w, v \rangle$ for all real numbers c and d,
 (iii) $\langle u, u \rangle \geq 0$,
 (iv) $\langle u, u \rangle = 0 \quad \Leftrightarrow \quad u \equiv 0$ (nontrivial).

In terms of the inner product, our penalty functional is

$$\mathcal{J}[u] = \langle u - u_F, u - u_F \rangle + (\mathbf{d} - \mathbf{u})^{\mathsf{T}} \mathbf{w} (\mathbf{d} - \mathbf{u}). \qquad (2.1.2)$$

□

2.1.3   Linear functionals and their representers; unobservables

Consider the linear mapping

$$u \to u(\xi, \tau), \qquad (2.1.3)$$

where the lhs is a field, while the rhs is a particular value of the field. This mapping is a *linear functional*: it linearly maps a function to a single number.

Theorem 2.1.1

If *the vector space of admissible fields u, with the inner product* $\langle \, , \, \rangle$*, is complete* (*that is, if it is a Hilbert Space*)*,* **then** *there is a function* $\rho(x, t, \xi, \tau)$ *such that*

$$\langle \rho, u \rangle = u(\xi, \tau). \qquad (2.1.4)$$

So ρ "represents" the measurement process. This is the Riesz representation theorem. Given ρ, we may express $\mathcal{J}$ entirely in terms of inner products (Wahba and Wendelberger, 1980):

$$\mathcal{J}[u] = \langle u - u_F, u - u_F \rangle + (\mathbf{d} - \langle \rho, u \rangle)^{\mathsf{T}} \mathbf{w} (\mathbf{d} - \langle \rho, u \rangle), \qquad (2.1.5)$$

where $\rho_m = \rho(x, t, x_m, t_m), 1 \le m \le M$.

Now, *any* field $u = u(x, t)$ may be expressed as

$$u(x, t) = u_F(x, t) + \sum_{m=1}^{M} v_m \rho(x, t, x_m, t_m) + g(x, t), \qquad (2.1.6)$$

where u_F is again the solution of (1.2.2)–(1.2.4), and where the v_m are *any* coefficients, since we may always choose

$$g \equiv u - u_F - \boldsymbol{\nu}^{\mathsf{T}} \rho. \qquad (2.1.7)$$

Let us now impose the condition that g is "unobservable":

$$\langle \rho_m, g \rangle = g(x_m, t_m) = 0 \qquad (2.1.8)$$

for $1 \le m \le M$. That is, g is orthogonal to each ρ_m. For a *given u* and a *given* u_F, we may use (2.1.8) to derive M equations for the v_m; then g is uniquely defined by (2.1.7).

2.1.4   Geometric minimization with representers

But we're *not* given u; we're only given u_F. Thus ν and g are arbitrary. We wish to find the u that minimizes $\mathcal{J}[u]$. Let us evaluate $\mathcal{J}$ using (2.1.5) and (2.1.6):

$$
\begin{aligned}
\mathcal{J}[u] &= \langle \nu^{\mathrm{T}}\rho + g, \nu^{\mathrm{T}}\rho + g \rangle \\
&\quad + (\mathbf{d} - \langle \rho, u_F + \rho^{\mathrm{T}}\nu + g \rangle)^{\mathrm{T}}\mathbf{w}(\mathbf{d} - \langle \rho, u_F + \rho^{\mathrm{T}}\nu + g \rangle) \\
&= \nu^{\mathrm{T}}\langle \rho, \rho^{\mathrm{T}} \rangle \nu + \nu^{\mathrm{T}}\langle \rho, g \rangle + \langle g, \rho^{\mathrm{T}} \rangle \nu + \langle g, g \rangle \\
&\quad + (\mathbf{d} - \langle \rho, u_F \rangle - \langle \rho, \rho^{\mathrm{T}} \rangle \nu - \langle \rho, g \rangle)^{\mathrm{T}}\mathbf{w}(\mathbf{d} - \langle \rho, u_F \rangle - \langle \rho, \rho^{\mathrm{T}} \rangle \nu - \langle \rho, g \rangle).
\end{aligned}
$$

$$(2.1.9)$$

Next impose the M orthogonality conditions (2.1.8), and use the representing property of ρ to obtain

$$
\begin{aligned}
\mathcal{J}[u] = \mathcal{J}[\nu, g] &= \nu^{\mathrm{T}}\langle \rho, \rho^{\mathrm{T}} \rangle \nu + \langle g, g \rangle \\
&\quad + (\mathbf{d} - \mathbf{u}_F - \langle \rho, \rho^{\mathrm{T}} \rangle \nu)^{\mathrm{T}}\mathbf{w}(\mathbf{d} - \mathbf{u}_F - \langle \rho, \rho^{\mathrm{T}} \rangle \nu).
\end{aligned}
$$

$$(2.1.10)$$

The penalty functional $\mathcal{J}$ is now expressed explicitly in terms of ν and g. Note that g only appears once on the rhs of (2.1.10). Clearly $\mathcal{J}$ is least with respect to the choice of g if $\langle g, g \rangle = 0$, that is

$$g = \hat{g} \equiv 0. \qquad (2.1.11)$$

We discard the field g orthogonal to all the representers. It remains to select the ν_m, $1 \le m \le M$. But first note that

$$
\begin{aligned}
\sigma_{lm} \equiv \langle \rho, \rho^{\mathrm{T}} \rangle_{lm} &= \langle \rho_l, \rho_m \rangle = \langle \rho_m, \rho_l \rangle \\
&= \rho_m(x_l, t_l) = \rho_l(x_m, t_m) \\
&= \sigma_{ml},
\end{aligned}
$$

$$(2.1.12)$$

so $\sigma = \sigma^{\mathrm{T}}$ and $\mathcal{J}$, which now only depends upon ν, may be expressed as

$$\mathcal{J}[u] = \mathcal{J}[\nu] = \nu^{\mathrm{T}}\sigma\nu + (\mathbf{h} - \sigma\nu)^{\mathrm{T}}\mathbf{w}(\mathbf{h} - \sigma\nu), \qquad (2.1.13)$$

where $\mathbf{h} \equiv \mathbf{d} - \mathbf{u}_F$. Completing the square,

$$\mathcal{J}[\nu] = (\nu - \hat{\nu})^{\mathrm{T}}\mathbf{S}(\nu - \hat{\nu}) + \mathbf{h}^{\mathrm{T}}\mathbf{w}\mathbf{h} - \hat{\nu}^{\mathrm{T}}\mathbf{S}\hat{\nu}, \qquad (2.1.14)$$

where $\mathbf{S} = \sigma + \sigma\mathbf{w}\sigma$ and $\mathbf{S}\hat{\nu} = \sigma\mathbf{w}\mathbf{h}$, both of which are given. We finally minimize $\mathcal{J}$ by choosing $\nu = \hat{\nu}$, and then $\hat{\mathcal{J}} \equiv \mathcal{J}[\hat{\nu}] = \mathbf{h}^{\mathrm{T}}\mathbf{w}\mathbf{h} - \hat{\nu}^{\mathrm{T}}\mathbf{S}\hat{\nu}$. Provided σ is nonsingular, we may untangle these results to find

$$(\sigma + \mathbf{w}^{-1})\hat{\nu} = \mathbf{h}, \qquad (2.1.15)$$

which looks familiar.

Recall that our minimizer is

$$\hat{u} = u_F + \hat{\nu}^{\mathrm{T}}\rho + \hat{g}, \qquad (2.1.16)$$

where $\hat{g}$ satisfies (2.1.11), and $\hat{\nu}$ satisfies (2.1.15).

2.1.5   Equivalence of variational and geometric minimization: the data space

Surely the representers ρ defined by the representing property (2.1.4) are the same as the representer functions $\mathbf{r}$ that satisfy the Euler–Lagrange-like system (1.3.8)–(1.3.13), in which case

$$\sigma = \mathbf{R}, \quad \hat{\nu} = \hat{\beta} ? \qquad (2.1.17)$$

Exercise 2.1.3
Show that

$$\rho_m(x, t) = r_m(x, t) \qquad (2.1.18)$$

for $0 \leq x \leq L, 0 \leq t \leq T$, and $1 \leq m \leq M$.

Hint
Consider (2.1.4):

$$
\begin{aligned}
u(x_m, t_m) &\equiv \langle \rho_m, u \rangle \\
&\equiv \int_0^T dt \int_0^T ds \int_0^L dx \int_0^L dy \left\{ \left(\frac{\partial}{\partial t} + c\frac{\partial}{\partial x} \right) \rho_m(x, t) \right\} \\
&\quad \times W_f(x, t, y, s) \left\{ \left(\frac{\partial}{\partial s} + c\frac{\partial}{\partial y} \right) \right\} u(y, s) \\
&\quad + \int_0^L dx \int_0^L dy \, \rho_m(x, 0) W_i(x, y) u(y, 0) \\
&\quad + \int_0^T dt \int_0^T ds \, \rho_m(0, t) W_b(t, s) u(0, s).
\end{aligned}
\qquad (2.1.19)
$$

Integrate the first integral by parts, and then compare (2.1.19) to

$$u(x_m, t_m) = \int_0^T dt \int_0^L dx \, u(x, t)\delta(x - x_m)\delta(t - t_m). \qquad (2.1.20)$$

Since the partially-integrated (2.1.19) must agree with (2.1.20) for all fields $u(x, t)$ having initial values $u(x, 0)$ and boundary values $u(0, t)$, we may equate their respective coefficients, arriving at (1.3.8)–(1.3.13). We have proved that for any field u,

$$\boxed{\langle \mathbf{r}, u \rangle = \mathbf{u}.} \qquad (2.1.21)$$

$\square$

Note 1. We have established that

$$\hat{u}(x, t) - u_F(x, t) = \hat{\boldsymbol{\beta}}^{\mathrm{T}} \mathbf{r}(x, t). \qquad (2.1.22)$$

That is, the difference between the inverse estimate $\hat{u}$ and the prior estimate u_F is a linear combination of the M representers $r_1, \ldots, r_M$. The difference lies in the *observable* space, that is, we reject any additional difference g that is *unobservable*: $\langle \mathbf{r}, g \rangle = 0$. We began with a search for the optimal or best-fit *field* $\hat{u}(x, t)$, where $0 \leq x \leq L$ and $0 \leq t \leq T$. This would be a search amongst an infinite number of degrees of freedom (the "state space"). We have exactly reduced the task to a search for the M optimal representer coefficients $\hat{\beta}_1, \ldots, \hat{\beta}_M$: see Fig. 2.1.1. These are the *observable* degrees of freedom (the "data space").

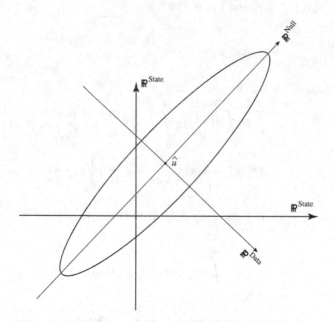

Figure 2.1.1 The plane represents the state space. It has an axis $u(x, t)$ for each (x, t) in the intervals $0 \leq x \leq L$, $0 \leq t \leq T$. In principle this is an infinite dimensional space. In practice, when we replace continuous intervals and partial differential equations with grids and partial difference equations, the state space usually has a very large but finite dimension. The contour is defined by a constant value for the penalty functional $\mathcal{J}[u]$, and has principal axes $\mathbb{R}^{\mathrm{Data}}$ (for $\boldsymbol{\beta}^{\mathrm{T}}\mathbf{r}(x, t)$) and $\mathbb{R}^{\mathrm{Null}}$ (for $g(x, t)$). Note that the representers $r_1(x, t), \ldots, r_M(x, t)$ are known, and *span* the data space, so only the unknown $\beta_1, \ldots, \beta_M$ vary in the data space. The unobservable field $g(x, t)$ is unknown and variable for $0 \leq x \leq L, 0 \leq t \leq T$. Realizing that $\hat{g}$ is zero greatly reduces the size of the search for $\hat{u}$, as we need only search in the data space.

Note 2. Recall from §1.3.3 the definition of Γ as the representer for point measurements. Hence for any field u,

$$\langle \Gamma, u \rangle = u, \tag{2.1.23}$$

and consequently Γ is known as a "reproducing kernel."

2.2 Statistical interpretation: the relationship to "optimal interpolation"

2.2.1 Random errors

Viewed as a generalized inverse or as a control problem, the ocean circulation u is estimated by adjusting the forcing f, initial value i, and boundary value b in order to obtain a better fit to the data, given that there are weights or "costs" W_f, W_i, W_b and $\mathbf{w}$ for these control variables and for the data misfit. Alternatively, the fields f, b and i for $0 \le x \le L$ and $0 \le t \le T$, and the data error vector ϵ, may be viewed as members of an *ensemble* of such quantities. That is, they are *random*. We shall attempt to estimate which f, i, b and ϵ were present in the "ocean" in $0 \le x \le L$ during our "cruise" for $0 \le t \le T$. We shall recover an interpretation of the "variational assimilation" of §1.2 in terms of the "optimal interpolation" routinely used in meteorology and oceanography.

2.2.2 Null hypotheses

In order to make these estimates, we shall have to make some assumptions about the ensemble. These assumptions compose a *null hypothesis* $\mathcal{H}_0$:

$$Ef(x,t) = Ei(x) = Eb(t) = E\epsilon_m = 0 \tag{2.2.1}$$

for $0 \le x \le L, 0 \le t \le T$ and $1 \le m \le M$, where $E(\ )$ denotes the ensemble average or mean;

$$E(f(x,t)f(y,s)) = C_f(x,t,y,s),$$
$$E(i(x)i(y)) = C_i(x,y),$$
$$E(b(t)b(s)) = C_b(t,s), \tag{2.2.2}$$

and

$$E(\epsilon\epsilon^{\mathrm{T}}) = \mathbf{C}_\epsilon,$$

while

$$E(fi) = E(fb) = E(f\epsilon_m) = E(ib) = E(i\epsilon_m) = E(b\epsilon_m) = 0. \tag{2.2.3}$$

Note 1. The covariances C_f, C_i, C_b and $\mathbf{C}_\epsilon$ are explicit functional or tabular forms.

Note 2. Only first and second moments are given in $\mathcal{H}_0 = \{(2.2.1), (2.2.2), (2.2.3)\}$; if the random variables f, i, b and ϵ are Gaussian, these moments determine the probability distribution function (pdf).

Note 3. The *alternative hypothesis* is that either (2.2.1), (2.2.2), or (2.2.3) is *not* true.

2.2.3   The reproducing kernel is a covariance

Let us now define $v = v(x, t)$ by

$$u(x, t) = u_F(x, t) + v(x, t). \tag{2.2.4}$$

That is, v is the random error in our prior estimate or forward solution u_F. The latter corresponds to our prior estimates F, I and B for the forcing, etc. These estimates are made *prior* to knowing the data $\mathbf{d}$. Clearly

$$\frac{\partial v}{\partial t} + c\frac{\partial v}{\partial x} = f, \tag{2.2.5}$$

subject to $v = i$ at $t = 0$, and $v = b$ at $x = 0$. We may use the Green's function γ to write

$$v = \gamma \bullet f + \gamma \circ i + c\gamma * b. \tag{2.2.6}$$

Hence

$$Ev = 0. \tag{2.2.7}$$

Exercise 2.2.1

Derive in detail the covariance for v:

$$C_v \equiv E(vv) = \gamma \bullet E(ff) \bullet \gamma + \gamma \circ E(ii) \circ \gamma + c^2\gamma * E(bb) * \gamma$$
$$= \gamma \bullet C_f \bullet \gamma + \gamma \circ C_i \circ \gamma + c^2\gamma * C_b * \gamma \tag{2.2.8}$$
$$= \Gamma. \tag{2.2.9}$$

That is, the covariance of the errors in the prior estimate is just the reproducing kernel (Weinert, 1982; Bennett, 1992).   □

2.2.4   "Optimal Interpolation", or best linear unbiased estimation; equivalence of generalized inversion and OI

We shall now outline the method of "optimal interpolation" (OI) for estimating a field u, given a first-guess u_F and data $\mathbf{d}$ (Bretherton *et al.*, 1976; Daley, 1991; Thiébaux and Pedder, 1987). The first guess need not be a model solution.

Suppose the true field is

$$u(x, t) = u_F(x, t) + q(x, t), \tag{2.2.10}$$

and suppose as before that

$$\mathbf{d} = \mathbf{u} + \boldsymbol{\epsilon}, \tag{2.2.11}$$

where

$$Eq(x, t) = E\epsilon_m = 0 \tag{2.2.12}$$

for $0 \le x \le L$, $0 \le t \le T$ and $1 \le m \le M$; and suppose that

$$\left. \begin{array}{l} E(q(x, t)q(y, s)) = C_q(x, t, y, s), \\ E(\boldsymbol{\epsilon\epsilon}^{\mathrm{T}}) \qquad\quad = \mathbf{C}_\epsilon, \\ \text{and} \\ E(q(x, t)\boldsymbol{\epsilon}) \qquad\;\; = \mathbf{0} \end{array} \right\} \tag{2.2.13}$$

for $0 \le x, y \le L$, $0 \le t, s \le T$.

Note 1. $E(\;)$ denotes an ensemble average, given the prior estimate u_F.
Note 2. $Eu = Eu_F + Eq = u_F + 0 = u_F$.
Note 3. $E\mathbf{d} = E\mathbf{u} + E\boldsymbol{\epsilon} = E\mathbf{u}_F + Eq + E\boldsymbol{\epsilon} = \mathbf{u}_F + \mathbf{0} + \mathbf{0} = \mathbf{u}_F$.

We seek the best linear unbiased estimate of u, that is

$$u_L(x, t) = u_F(x, t) + (\mathbf{d} - \mathbf{u}_F)^{\mathrm{T}}\mathbf{s}(x, t), \tag{2.2.14}$$

where $s_1(x, t), \ldots, s_M(x, t)$ are M as yet unchosen non-random interpolants.

Note 1. u_L is linear in u_F and $\mathbf{d}$.
Note 2. u_L is unbiased: $Eu_L = u_F = Eu$.
Note 3. u_L is best if

$$Ee_L^2(x, t) \equiv E\{u(x, t) - u_L(x, t)\}^2 \tag{2.2.15}$$

is least for *each* (x, t).

Now

$$Ee_L^2 = E\{u_F + q - u_F - (\mathbf{u}_F + \mathbf{q} + \boldsymbol{\epsilon} - \mathbf{u}_F)^{\mathrm{T}}\mathbf{s}\}^2 \tag{2.2.16}$$

$$= E\{q - (\mathbf{q} + \boldsymbol{\epsilon})^{\mathrm{T}}\mathbf{s}\}^2$$

$$= Eq^2 - E\{q(\mathbf{q} + \boldsymbol{\epsilon})^{\mathrm{T}}\}\mathbf{s} - \mathbf{s}^{\mathrm{T}} E\{(\mathbf{q} + \boldsymbol{\epsilon})q\}$$

$$+ \mathbf{s}^{\mathrm{T}} E\{(\mathbf{q} + \boldsymbol{\epsilon})(\mathbf{q} + \boldsymbol{\epsilon})^{\mathrm{T}}\}\mathbf{s} \tag{2.2.17}$$

for each (x, t). We want

$$\frac{\partial Ee_L^2}{\partial s_m} = 0 \tag{2.2.18}$$

for $1 \leq m \leq M$, at each (x, t). That is,

$$-E\{(\mathbf{q} + \boldsymbol{\epsilon})q\} + E\{(\mathbf{q} + \boldsymbol{\epsilon})(\mathbf{q} + \boldsymbol{\epsilon})^{\mathrm{T}}\}\mathbf{s} = \mathbf{0},$$

or

$$-E(\mathbf{q}q) + \{E(\mathbf{q}\mathbf{q}^{\mathrm{T}}) + E(\boldsymbol{\epsilon}\boldsymbol{\epsilon}^{\mathrm{T}})\}\mathbf{s} = \mathbf{0} \qquad (2.2.19)$$

since $E(q\boldsymbol{\epsilon}) = \mathbf{0}$ by assumption. In detail, (2.2.19) is

$$-C_q(x, t, x_n, t_n) + \sum_{m=1}^{M} \{C_q(x_n, t_n, x_m, t_m) + C_{\epsilon_{n,m}}\}s_m(x, t) = 0 \qquad (2.2.20)$$

for $0 \leq x \leq L, 0 \leq t \leq T$ and $1 \leq n \leq M$. Solving (2.2.20) for s_n yields

$$s_n(x, t) = \sum_{m=1}^{M} \{\mathbf{C}_q + \mathbf{C}_\epsilon\}_{n,m}^{-1} C_q(x, t, x_m, t_m), \qquad (2.2.21)$$

where the superscript "-1" indicates a matrix inverse, and $\{\mathbf{C}_q\}_{n,m} = C_q(x_n, t_n, x_m, t_m)$. These $\mathbf{s}(x, t)$ are the optimal interpolants. They do not depend upon u_F or $\mathbf{d}$, but do depend upon the prior covariances C_q and $\mathbf{C}_\epsilon$. In conclusion our best linear unbiased estimate or BLUE is

$$u_L(x, t) = u_F(x, t) + (\mathbf{d} - \mathbf{u}_F)^{\mathrm{T}} \{\mathbf{C}_q + \mathbf{C}_\epsilon\}^{-1} C_q(x, t). \qquad (2.2.22)$$

Now compare (2.2.22) with (1.3.24), and recall the first line in (1.3.32).

We have proved:

> Generalized inversion (the minimization of the integral penalty functional $\mathcal{J}[u]$) is the same as optimal interpolation (the minimization of the local error variance $Ee_L^2(x, t)$) when the solution of the forward model u_F is the mean field, when the data weight matrix $\mathbf{w}$ is the inverse of the data error covariance matrix $\mathbf{C}_\epsilon$, and when the reproducing kernel Γ is the covariance C_q. In particular (see (2.2.8), (2.2.9)),

$$r_m(x, t) = \Gamma(x, t, x_m, t_m) = C_v(x, t, x_m, t_m) = C_q(x, t, x_m, t_m),$$
$$R_{nm} = \Gamma(x_n, t_n, x_m, t_m) = C_v(x_n, t_n, x_m, t_m) = C_q(x_n, t_n, x_m, t_m).$$

Generalized Inversion is Optimal Interpolation

Note 1. Our model is linear, and the data are pointwise.
Note 2. OI is widely used in meteorology and oceanography, for the "analysis" or "mapping" of scalar data when the statistical properties of the fields are plausibly independent of coordinate origins or orientations, both in space and time. That is, when

$$C_q(x, t, y, s) = C_q(|x - y|, |t - s|), \qquad (2.2.23)$$

for example. Such covariances need only involve a few parameters, which should be reliably estimable from reasonably large sets of data. However, we are increasingly obliged to admit that different fields are dependent, on dynamical or chemical or biological grounds, so we should use multivariate or vector forms of OI. Moreover, planetary-scale and coastal circulation are obviously statistically inhomogeneous, while the endless emergence of trends suggests statistical nonstationarity. That is, (2.2.23) is false. OI may be generalized to the multivariate, inhomogeneous and nonstationary case provided that there are credible prior estimates for all the parameters in the covariances of the fields being mapped. We hope that our dynamical models are getting so faithful to the larger scales that model errors like f must be limited to the smaller scales at which (2.2.23) may be plausible. Thus, we should only need to estimate $C_f(x, t, y, s) = C_f(|x - y|, |t - s|)$. We may then use generalized inversion to generate, in effect, the inhomogeneous and nonstationary multivariate equivalents of $C_v = C_v(x, t, y, s)$ and then perform, in effect, an OI of the data.

A serious caution must now be offered. It is misleadingly easy to declare that the dynamical error f, initial error i, etc., are random variables belonging to some ensemble, and to manipulate their ensemble moments Ef, Ei, $E(ff)$, $E(fi)$, etc. It is much harder to devise a credible method for estimating these moments. The fields must clearly be statistically homogeneous at least in one spatial direction or in time, but the presence of spatial or climatological trends makes such homogeneity far from clear. Worse, our dynamical models have already been Reynolds-averaged or subgridscale-averaged, so f in particular is already an average of a certain kind. The statistical interpretation of variational assimilation requires, therefore, a *second* randomization. This difficult issue will be discussed in greater detail in §5.3.7.

2.3   The reduced penalty functional

2.3.1   Inversion as hypothesis testing

Inverse methods enable us to smooth data using a dynamical model as a constraint. Equally, the methods enable us to test the model using the data. The concept of a model is extended here to include not only equations of motion, initial conditions and boundary conditions, but also an hypothesis concerning the errors in each such piece of information. If the model fails the test for a given data set, then the interpolated data or "analysis field" is suspect. If the test is failed repeatedly for many data sets, then the hypothesis is suspect. This would be an unsatisfactory state of affairs from the point of view of the ocean analyst or ocean forecaster, but should please the ocean modeler: something new would have been learned about the ocean, namely, that the errors in the dynamics, initial conditions or boundary conditions had been underestimated. Lagrange multipliers make it possible (exercise!) to distinguish between forcing errors and additive components of parameterization errors.

The hypothesis test is based on the statistical interpretation developed in the previous section. The derivation of the test is very short, but the realization that inverse methods enable model testing is so important that a separate section is warranted.

2.3.2   Explicit expression for the reduced penalty functional

First, recall from (1.3.26) and §2.2.3 that the minimum value of the penalty functional $\mathcal{J}$ is

$$\hat{\mathcal{J}} \equiv \mathcal{J}[\hat{u}] = \mathbf{h}^{\mathrm{T}}\mathbf{P}^{-1}\mathbf{h}$$

$$= (\mathbf{d} - \mathbf{u}_F)^{\mathrm{T}}(\mathbf{R} + \mathbf{C}_\epsilon)^{-1}(\mathbf{d} - \mathbf{u}_F), \qquad (2.3.1)$$

where the actual or true "ocean circulation" is

$$u = u_F + v, \qquad (2.3.2)$$

where v is the model response to the random inputs f, i and b, and the data are

$$\mathbf{d} = \mathbf{u} + \boldsymbol{\epsilon}, \qquad (2.3.3)$$

where ϵ is random measurement error.

Hence

$$\mathbf{h} \equiv \mathbf{d} - \mathbf{u}_F = \mathbf{u} + \boldsymbol{\epsilon} - (\mathbf{u} - \mathbf{v}) = \boldsymbol{\epsilon} + \mathbf{v}, \qquad (2.3.4)$$

$$E\mathbf{h} = E\boldsymbol{\epsilon} + E\mathbf{v} = \mathbf{0}, \qquad (2.3.5)$$

$$E(\mathbf{h}\mathbf{h}^{\mathrm{T}}) = E((\boldsymbol{\epsilon} + \mathbf{v})(\boldsymbol{\epsilon} + \mathbf{v})^{\mathrm{T}})$$

$$= E(\boldsymbol{\epsilon}\boldsymbol{\epsilon}^{\mathrm{T}}) + E(\boldsymbol{\epsilon}\mathbf{v}^{\mathrm{T}}) + E(\mathbf{v}\boldsymbol{\epsilon}^{\mathrm{T}}) + (\mathbf{v}\mathbf{v}^{\mathrm{T}})$$

$$= \mathbf{C}_\epsilon + \mathbf{0} + \mathbf{0} + \mathbf{R} = \mathbf{P} + E(\mathbf{v}\mathbf{v}^{\mathrm{T}}), \qquad (2.3.6)$$

which statements are parts of, or consequences of, our null hypothesis $\mathcal{H}_0$ defined by (2.2.1)–(2.2.3).

Now define $\mathbf{P}^{\frac{1}{2}}$, which is meaningful since $\mathbf{P}$ is positive-definite and symmetric, and hence define

$$\mathbf{k} \equiv \mathbf{P}^{-\frac{1}{2}}\mathbf{h}. \qquad (2.3.7)$$

Then

$$E\mathbf{k} = \mathbf{P}^{-\frac{1}{2}} E\mathbf{h} = \mathbf{0}, \qquad (2.3.8)$$

$$E(\mathbf{k}\mathbf{k}^{\mathrm{T}}) = \mathbf{P}^{-\frac{1}{2}} E(\mathbf{h}\mathbf{h}^{\mathrm{T}}) \, \mathbf{P}^{-\frac{1}{2}}$$

$$= \mathbf{P}^{-\frac{1}{2}} \mathbf{P}\mathbf{P}^{-\frac{1}{2}}$$

$$= \mathbf{I}. \qquad (2.3.9)$$

That is,

$$E(k_n k_m) = \delta_{nm}. \qquad (2.3.10)$$

2.3.3   Statistics of the reduced penalty: χ^2 testing

The scaled, prior data misfits $k_1, \ldots, k_M$ are zero-mean, uncorrelated, unit-variance random variables and

$$
\begin{aligned}
\hat{\mathcal{J}} = \mathbf{h}^{\mathrm{T}}\mathbf{P}^{-1}\mathbf{h} &= \mathbf{k}^{\mathrm{T}}\mathbf{P}^{\frac{1}{2}}\,\mathbf{P}^{-1}\mathbf{P}^{\frac{1}{2}}\,\mathbf{k} \\
&= \mathbf{k}^{\mathrm{T}}\mathbf{k} \\
&= k_1^2 + \cdots + k_M^2.
\end{aligned}
\tag{2.3.11}
$$

Therefore $\hat{\mathcal{J}} = \chi_M^2$, the chi-squared random variable with M degrees of freedom. Or is it? To be precise,

$$
\chi_M^2 = x_1^2 + \cdots + \cdots + x_M^2,
\tag{2.3.12}
$$

where the pdf for x_m is

$$
p(x_m) = (2\pi)^{-\frac{1}{2}}\exp\left(-x_m^2/2\right),
\tag{2.3.13}
$$

$1 \le m \le M$. That is, each x_m is a Gaussian random variable having zero mean and unit variance: $x_m \sim N(0, 1)$ (Press *et al.*, 1986). If we had included in $\mathcal{H}_0$ the assumption that f, i, b and ϵ were Gaussian, then by linearity $\mathbf{h}$ and hence $\mathbf{k}$ would also be Gaussian. If we do not make that assumption, we may invoke the *central limit theorem* when M is large, to infer that

$$
k_n = \sum_{m=1}^{M}(\mathbf{P}^{-\frac{1}{2}})_{nm}h_m \sim N(0, 1)
\tag{2.3.14}
$$

as $M \to \infty$.

But roughly, if $\mathcal{H}_0$ is true then

$$
\hat{\mathcal{J}} = \chi_M^2.
\tag{2.3.15}
$$

So we have a chi-squared test for our null hypothesis. Now

$$
E\left(\chi_M^2\right) = M, \quad \mathrm{var}\left(\chi_M^2\right) \equiv E\left((\chi_M^2)^2\right) - \left(E\left(\chi_M^2\right)\right)^2 = 2M.
\tag{2.3.16}
$$

If we perform the inversion a number of times with different data, and find that our sample distribution has significantly bigger first or second moments than those of χ_M^2, then we should *reject* $\mathcal{H}_0$. *We would have learned something about the ocean, from the data.* Specifically, we would have learned that the ocean differs from the model, by *more* than we had hypothesized. Recall that $\mathcal{J}$ is inversely proportional to C_f, etc.

Exercise 2.3.1
Show that

(i) $\mathcal{J}_F \equiv \mathcal{J}[u_F] = \mathbf{h}^{\mathrm{T}}\mathbf{C}_{\epsilon}^{-1}\,\mathbf{h},$
$\qquad\qquad\qquad\qquad\qquad\qquad\qquad\qquad\qquad$ (2.3.17)

(ii)　$\mathcal{J}_{\text{mod}} \equiv \langle \hat{u} - u_F, \hat{u} - u_F \rangle = \hat{\beta}^{\text{T}} \mathbf{R} \hat{\beta},$　(2.3.18)

(iii)　$\mathcal{J}_{\text{data}} \equiv (\hat{u} - \mathbf{d})^{\text{T}} \mathbf{C}_\epsilon^{-1} (\hat{u} - \mathbf{d}) = \hat{\beta}^{\text{T}} \mathbf{C}_\epsilon \hat{\beta},$　(2.3.19)

where $\mathbf{P}\hat{\beta} = \mathbf{h}$.

Note 1. $\mathcal{J}_F$ is "only data misfit".
Note 2. In general, $\hat{\mathcal{J}}_{\text{mod}} \neq \hat{\mathcal{J}}_{\text{data}}$.　□

Exercise 2.3.2
Show that

(i)　$E\mathcal{J}_F = \text{Tr}\left(\mathbf{C}_\epsilon^{-\frac{1}{2}} \mathbf{R} \mathbf{C}_\epsilon^{-\frac{1}{2}}\right) + M,$　(2.3.20)

(ii)　$E\hat{\mathcal{J}}_{\text{mod}} = \text{Tr}\left(\mathbf{R}^{\frac{1}{2}} \mathbf{P}^{-1} \mathbf{R}^{\frac{1}{2}}\right),$　(2.3.21)

(iii)　$E\hat{\mathcal{J}}_{\text{data}} = \text{Tr}\left(\mathbf{C}_\epsilon^{\frac{1}{2}} \mathbf{P}^{-1} \mathbf{C}_\epsilon^{\frac{1}{2}}\right).$　(2.3.22)

Note 1. Usually $E\mathcal{J}_F \gg E\hat{\mathcal{J}} = M$.
Note 2. In general, $E\hat{\mathcal{J}}_{\text{mod}} \neq E\hat{\mathcal{J}}_{\text{data}}$.
Note 3. In order to devise a rigorous and objective test for an ocean model, we have extended the definition of a model to include an hypothesis about the statistics of the errors in the dynamics, in the initial conditions and in the boundary conditions as well as in the data.　□

Exercise 2.3.3
Give meanings to the left-hand sides of (2.3.17)–(2.3.22).　□

Exercise 2.3.4 (Bennett et al., 2000)
Show that

(i)　$\text{var}\,(\mathcal{J}_F) \sim 2\,\text{Tr}\,(\mathbf{C}_\epsilon^{-1} \mathbf{P}^2 \mathbf{C}_\epsilon^{-1}),$　(2.3.23)

(ii)　$\text{var}\,(\hat{\mathcal{J}}_{\text{data}}) \sim 2\,\text{Tr}\,(\mathbf{C}_\epsilon \mathbf{P}^{-2} \mathbf{C}_\epsilon),$　(2.3.24)

(iii)　$\text{var}\,(\hat{\mathcal{J}}_{\text{mod}}) \sim 2\,\text{Tr}\,[(\mathbf{I} - \mathbf{P}^{-\frac{1}{2}} \mathbf{C}_\epsilon \mathbf{P}^{-\frac{1}{2}})^2]$　(2.3.25)

as $M \to \infty$.　□

Remark
It is difficult to develop a credible null hypothesis $\mathcal{H}_0$. In particular it is difficult to develop the covariances C_f, etc. It follows that $\hat{u}$, the resulting inverse estimate or analysis of the circulation, also lacks credibility. It is a misconception, however, to view inversion as "garbage in, garbage out". Rather, inversion puts the hypothesis to the test. Forward modeling is no less exposed to the charge of "garbage in, garbage

out": it tests the nullest of null hypotheses, namely, that the dynamical errors, initial errors and boundary errors are all zero, which is the rankest of garbage.

2.4   General measurement

2.4.1   Point measurements

Our data thus far have been direct measurements of the circulation field u at isolated points in space and time:

$$d_m = u(x_m, t_m) + \epsilon_m \qquad (2.4.1)$$

for $1 \le m \le M$, where d_m is the datum and ϵ_m the measurement error. We shall now consider more general measurements (Bennett, 1985, 1990).

2.4.2   Measurement functionals

First note that a map sending a *field* u into a *single real number* $u(z, w)$ is an example of a *functional*

$$u \to \mathcal{L}[u] = u(z, w). \qquad (2.4.2)$$

In general u may be a *vector field* of velocity components, pressure, temperature etc., but a measurement of u produces a *single number*. If the field is a streamfunction ψ, and the datum is the meridional component of velocity collected from a current meter, then the appropriate functional is

$$\psi \to \frac{\partial \psi}{\partial x}(z, w); \qquad (2.4.3)$$

if the field is sea-level elevation h, and the datum is the vertical acceleration of a wave-rider buoy, then

$$h \to \frac{\partial^2 h}{\partial t^2}(z, w); \qquad (2.4.4)$$

if the field is fluid velocity u along a zonal acoustic path, and the datum derives from reciprocal-shooting tomography, then

$$u \to \int_{z_1}^{z_2} u(x, w)\,dx; \qquad (2.4.5)$$

if the field is sea-level elevation, and the datum is collected by a radar beam, then

$$h \to \int_0^T dt \int_0^L dx\, K(x, t)h(x, t); \qquad (2.4.6)$$

finally, if the field is stratospheric temperature θ and the datum is a radiative energy flux, then by Stefan's law,

$$\theta \to \theta^4(z, w). \tag{2.4.7}$$

Note 1. Examples (2.4.2)–(2.4.6) are *linear*:

$$\mathcal{L}[au + bv] = a\mathcal{L}[u] + b\mathcal{L}[v] \tag{2.4.8}$$

for any fields u, v and real numbers a, b.

Note 2. Example (2.4.7) is *nonlinear*.

Note 3. Each of (2.4.2)–(2.4.5) can be expressed as (2.4.6):

$$K(x, t) = \delta(x - z)\delta(t - w) \quad \text{for} \quad (2.4.2),$$
$$K(x, t) = -\delta'(x - z)\delta(t - w) \quad \text{for} \quad (2.4.3),$$
$$K(x, t) = \delta(x - z)\delta''(t - w) \quad \text{for} \quad (2.4.4).$$

Exercise 2.4.1

Find K for (2.4.5). □

2.4.3   Representers for linear measurement functionals

The penalty functional may be now expressed as

$$\mathcal{J}[u] = \langle u - u_F, u - u_F \rangle + (\mathbf{d} - \mathcal{L}[u])^{\mathrm{T}} \mathbf{C}_\epsilon^{-1} (\mathbf{d} - \mathcal{L}[u]), \tag{2.4.9}$$

where $\mathcal{L}^{\mathrm{T}} = (\mathcal{L}_1, \ldots, \mathcal{L}_M)$ indicates M linear measurement functionals. Note that in earlier sections,

$$\mathcal{L}_m[u] \equiv u(x_m, t_m). \tag{2.4.10}$$

Furthermore, the Riesz representation theorem establishes that if $u \to \mathcal{L}_m[u]$ is a *bounded* ($\sup |\mathcal{L}_m[u]| / \| u \| < \infty$) linear functional acting on a Hilbert space, then there is an element (a field) r_m in the space such that

$$\langle u, r_m \rangle = \mathcal{L}_m[u] \tag{2.4.11}$$

for any field u.

Exercise 2.4.2

Verify that

$$r_m(x, t) = \mathcal{L}_{m_{(y,s)}}[\Gamma(x, t, y, s)], \tag{2.4.12}$$

where Γ is the reproducing kernel, and the subscripts (y, s) indicate that $\mathcal{L}_m$ acts on Γ as a field over (y, s), for each (x, t). Recall that Γ is the representer for evaluation of u at (y, s). In fact, show that r_m satisfies

$$\frac{\partial r_m}{\partial t} + c\frac{\partial r_m}{\partial x} = C_f \bullet \alpha_m, \qquad (2.4.13)$$

subject to $r_m = C_i \circ \alpha_m$ at $t = 0$, and $r_m = cC_b * \alpha_m$ at $x = 0$, where

$$-\frac{\partial \alpha_m}{\partial t} - c\frac{\partial \alpha_m}{\partial x} = \mathcal{L}_{m_{(y,s)}}[\delta(x - y)\delta(t - s)], \qquad (2.4.14)$$

subject to $\alpha_m = 0$ at $t = T$, and $\alpha_m = 0$ at $x = L$. $\qquad\square$

We may now write

$$\mathcal{J}[u] = \langle u - u_F, u - u_F \rangle + (\mathbf{d} - \langle u, \mathbf{r} \rangle)^\mathrm{T} \mathbf{C}_\epsilon^{-1}(\mathbf{d} - \langle u, \mathbf{r} \rangle) \qquad (2.4.15)$$

and, as before, the minimizer is

$$\hat{u} = u_F + \mathbf{h}^\mathrm{T}\mathbf{P}^{-1}\mathbf{r}, \qquad (2.4.16)$$

where

$$\mathbf{h} \equiv \mathbf{d} - \mathcal{L}[u_F] \qquad (2.4.17)$$

and $\mathbf{P} = \mathbf{R} + \mathbf{C}_\epsilon$, where

$$\mathbf{R} = \langle \mathbf{r}, \mathbf{r}^\mathrm{T} \rangle = \mathcal{L}[\mathbf{r}^\mathrm{T}] = \text{``}\mathcal{L}[\Gamma]\mathcal{L}^\mathrm{T}\text{''}. \qquad (2.4.18)$$

From now on we shall assume that $\mathbf{r}$, with its adjoint field α, represents a general linear measurement functional. We shall reserve the notation and nomenclature of the rk $\Gamma = \Gamma(x, t, z, w)$, with its adjoint field the Green's function $\gamma = \gamma(x, t, z, w)$, for the *evaluation functional* (2.4.2).

Exercise 2.4.3
Derive the Euler–Lagrange equations for extrema of (2.4.9). In particular, show that the generalization of (1.3.1) is

$$-\frac{\partial \lambda}{\partial t} - c\frac{\partial \lambda}{\partial x} = \mathcal{L}^\mathrm{T}[\delta\delta]\mathbf{C}_\epsilon^{-1}(\mathbf{d} - \mathcal{L}[\hat{u}]). \qquad (2.4.19)$$

$\qquad\square$

2.5   Array modes

2.5.1   Stable combinations of representers

We have seen that, amongst all free and forced solutions of the forward model, the observing system or "array" only detects the representers. We now ask: are some combinations of representers more stably detected than others?

2.5.2   Spectral decomposition, rotated representers

Assume general linear measurement functionals $\mathcal{L} = (\mathcal{L}_1, \ldots, \mathcal{L}_M)^{\mathrm{T}} : u \to \mathcal{L}[u] \in \mathbb{R}^M$. That is, $\mathcal{L}$ maps the field u linearly into the M real numbers $\mathcal{L}[u]$. The data $\mathbf{d}$ are of the form $\mathbf{d} = \mathcal{L}[u] + \epsilon$, where ϵ is the vector of measurement errors. The representer matrix is

$$R_{nm} = \mathcal{L}_{n_{(x,t)}} \mathcal{L}_{m_{(y,s)}} [\Gamma(x, t, y, s)] \tag{2.5.1}$$

for $1 \leq n, m \leq M$, where $\mathcal{L}_{n_{(x,t)}}$ acts on $\Gamma(x, t, \ , \ )$, etc. In vector notation,

$$\mathbf{R} = \mathcal{L}\Gamma\mathcal{L}^{\mathrm{T}}. \tag{2.5.2}$$

Recall again that the reproducing kernel Γ is also the covariance C_v: see §2.2, Exercise 2.2.1. The minimization of the penalty functional $\mathcal{J}$, defined by (1.5.7), reduces to the solution of the M-dimensional linear system

$$\mathbf{P}\hat{\beta} \equiv (\mathbf{R} + \mathbf{C}_\epsilon)\hat{\beta} = \mathbf{h} \equiv \mathbf{d} - \mathcal{L}[u_F], \tag{2.5.3}$$

where u_F is the solution of the forward model. The representer matrix $\mathbf{R}$ depends upon the dynamics, the prior covariances C_f, C_i and C_b for dynamical, initial and boundary residuals, and upon the array $\mathcal{L}$, while $\mathbf{C}_\epsilon$ is the covariance of measurement errors. Thus $\mathbf{P}$ encapsulates all of our prior knowledge of the ocean in general but does not depend upon the prior estimates of forcing, initial and boundary values F, B and I, provided the dynamics and measurement functionals are linear. The symmetry and positive definiteness of $\mathbf{P}$ implies the spectral decomposition

$$\mathbf{P} = \mathbf{Z}\Phi\mathbf{Z}^{\mathrm{T}}, \tag{2.5.4}$$

where $\mathbf{Z}$ is orthogonal: $\mathbf{Z}\mathbf{Z}^{\mathrm{T}} = \mathbf{Z}^{\mathrm{T}}\mathbf{Z} = \mathbf{I}$, and Φ is diagonal: $\Phi = \mathrm{diag}\,(\phi_1, \ldots, \phi_M)$, where $\phi_1 \geq \cdots \geq \phi_M > 0$.

Let $\mathcal{L}'$ be a rotated vector of measurement functionals:

$$\mathcal{L}' \equiv \mathbf{Z}^{\mathrm{T}}\mathcal{L}, \tag{2.5.5}$$

and define the rotated representers $\mathbf{r}' = \mathbf{r}'(x, t)$ by

$$\mathbf{r}' \equiv \mathbf{Z}^{\mathrm{T}}\mathbf{r} = \mathbf{Z}^{\mathrm{T}}\mathcal{L}[\Gamma]. \tag{2.5.6}$$

These are the *array modes* (Bennett, 1985, 1992). In particular,

$$\mathbf{R}' = \mathcal{L}'\Gamma\mathcal{L}'^{\mathrm{T}} = \mathbf{Z}^{\mathrm{T}}\mathcal{L}\Gamma\mathcal{L}^{\mathrm{T}}\mathbf{Z}$$
$$= \mathbf{Z}^{\mathrm{T}}\mathbf{R}\mathbf{Z}, \tag{2.5.7}$$

while

$$\mathbf{C}'_\epsilon = \mathbf{Z}^{\mathrm{T}}\mathbf{C}_\epsilon\mathbf{Z}. \tag{2.5.8}$$

Hence

$$\mathbf{P}' = \mathbf{R}' + \mathbf{C}'_\epsilon = \mathbf{Z}^{\mathrm{T}}\mathbf{P}\mathbf{Z} = \mathbf{\Phi}, \tag{2.5.9}$$

which is diagonal. The rotated representer coefficients $\hat{\beta}'$ then obey

$$\mathbf{P}'\hat{\beta}' = \mathbf{h}', \tag{2.5.10}$$

where

$$\mathbf{h}' = \mathbf{Z}^{\mathrm{T}}\mathbf{h}. \tag{2.5.11}$$

2.5.3   Statistical stability, clipping the spectrum

The solution for $\hat{\beta}'$ is trivial, since $\mathbf{P}' = \mathbf{\Phi}$ is diagonal:

$$\hat{\beta}'_m = \frac{h'_m}{\phi_m} \tag{2.5.12}$$

for $1 \le m \le M$. We may deduce from (2.3.5) and (2.3.6) that

$$E\mathbf{h}' = \mathbf{0}, \ E(\mathbf{h}'\mathbf{h}'^{\mathrm{T}}) = \mathbf{P}' = \mathbf{\Phi}, \tag{2.5.13}$$

so

$$E\hat{\beta}'_m = 0, \quad E((\hat{\beta}'_m)^2) = \frac{E((h'_m)^2)}{\phi_m^2} = \frac{\phi_m}{\phi_m^2} = \phi_m^{-1}. \tag{2.5.14}$$

That is, the estimated array mode coefficients $\hat{\beta}'_m$ have greater variance if the corresponding eigenvalue ϕ_m is smaller; (2.5.12) shows the inverse to be unstable if the prior data misfit $\mathbf{h}$ projects significantly onto eigenvectors of $\mathbf{P}$ having very small eigenvalues. Such projections should be discarded for $m > m_c$, where m_c is some cut-off. The exact inverse is

$$\hat{u} = u_F + \mathbf{r}^{\mathrm{T}}\hat{\beta} = u_F + \mathbf{r}^{\mathrm{T}}\mathbf{Z}\mathbf{Z}^{\mathrm{T}}\hat{\beta} = u_F + \mathbf{r}'^{\mathrm{T}}\hat{\beta}' = u_F + \mathbf{r}'^{\mathrm{T}}\mathbf{\Phi}^{-1}\mathbf{h}', \tag{2.5.15}$$

or

$$\hat{u}(x,t) = u_F(x,t) + \sum_{m=1}^{M} r'_m(x,t)\phi_m^{-1}h'_m, \tag{2.5.16}$$

and so the stabilized approximation is

$$\hat{u}(x, t) \cong u_F(x, t) + \sum_{m=1}^{m_c} r'_m(x, t)\phi_m^{-1}h'_m. \tag{2.5.17}$$

Array modes $r'_{m_c+1}, \ldots, r'_M$ have been made redundant.

Note 1. The components of the vector of rotated measurement functionals need not correspond to individual elements in the array. They correspond to linear combinations of the elements.

Note 2. If we arbitrarily make $\mathbf{P}$ more diagonally dominant:

$$\mathbf{P} \rightarrow \mathbf{P} + \sigma^2\mathbf{I}, \tag{2.5.18}$$

where σ^2 is additional, independent measurement error variance, then the eigenvalues of $\mathbf{P}$ become $\phi_1 + \sigma^2, \ldots, \phi_M + \sigma^2$, which all exceed σ^2. Thus (2.5.18) would seem to stabilize the inverse. Spectral decomposition (2.5.4), rotation (2.5.6) or clipping (2.5.17) would not be required. However, $\mathbf{P}$ and $\mathbf{P} + \sigma^2\mathbf{I}$ have the same eigenvectors, so the array modes are unaffected by (2.5.18). The modes $r'_{m_c+1}(x, t), \ldots, r'_M(x, t)$ usually have very fine structure, and retaining them at almost any level yields a "noisy" inverse $\hat{u}(x, t)$. It is better to clip the spectrum of $\mathbf{P}$ than to make $\mathbf{P}$ more diagonally dominant.

Note 3. The construction of array modes is essentially an analysis of the condition or stability of the generalized inverse of the model plus array, that is, the stability of the minimization of the penalty functional denoted by (1.5.7), (1.5.9) or (2.1.2). There are two major steps in constructing the inverse. The first is the discard (2.1.11) of all the unobservable fields (2.1.8); it is effected by admitting only solutions of the Euler–Lagrange equations. The second step is the solution of the finite-dimensional linear system denoted as (2.1.15) or (2.5.3). Once this system is solved, the coupling in the Euler–Lagrange equations is resolved and the generalized inverse is finally obtained by the explicit assembly of (1.3.24), or equivalently by a backward integration followed by a forward integration (see §3.1.2). The dimension of the algebraic system is M, the total number of data. The condition of the system is determined by the M eigenvalues of the coefficient matrix $\mathbf{P} = \mathbf{R} + \mathbf{C}_\epsilon$. The essential point is that the condition of the inverse is determined without first making a numerical approximation to the model using, say, finite differences; the condition is determined at the continuum level. That is, the condition is set by the partial differential equations, initial conditions, and boundary conditions of the model, by the measurement functionals for the observing system or array, and by the form and weighting of the penalty functional (the actual inputs to the model: internal forcing, initial values, boundary values and data values, have no influence; the stability of the inverse is its sensitivity to them as a class). The inevitable numerical approximation will indeed modify the null space of unobservable fields somewhat, and will also alter the eigenvalues, especially the smallest, but these effects are spurious and are suppressed in practice by physical diffusion in the dynamics, by convolution with the covariances in the Euler–Lagrange equations, and by the measurement error variance which has a stabilizing influence in general. Nevertheless, the continuum and discrete analyses of condition make for an interesting comparison. They may be found in Bennett (1985) and Courtier *et al.* (1993), respectively.

To end with a caution, it is imperative to realize that the array modes and assessment of conditioning depend not only upon the dynamics of the ocean model and the structure of the observing system or array, but also upon the hypothesized or prior covariances of the errors in the model and observing system. If subsequent testing of the hypothesis, using data collected by the array, leads to a rejection of the hypothesis, then the array assessment must also be rejected. Model testing and array assessment are inextricably intertwined. Examples will be presented in Chapter 5. For another approach to array design, see Hackert *et al.* (1998).

2.6   Smoothing norms, covariances and convolutions

2.6.1   Interpolation theory

The mathematical theory of interpolation is very old. It attracted the attention of the founders of analysis, including Newton, Lagrange and Gauss. The subject was in an advanced state of development by 1940; it then experienced a major reinvigoration with the advent of electronic computers. See Press *et al.* (1986; Section 2) for a neat outline of common methods, and Daley (1991; Chapter 2) for an authoritative account of methods widely used in meteorology and oceanography. What follows here is a brief outline of the theory attributed to E. Parzen, linking analytical and statistical interpolation. Aside from offering deeper insight into penalty functionals, the theory enables us to design and "tune" roughness penalties essentially equivalent to prescribed covariances (and vice versa). This is of critical importance if one intends, either out of taste or necessity, to minimize a penalty functional by searching in the control subspace rather than in the data subspace. The former search requires roughness penalties or weighting operators; the latter search exploits the Euler–Lagrange equations which incorporate covariances.

It has been argued in §2.1 that the data-subspace search is in principle highly efficient, but this efficiency will be wasted if the convolution-like integrals of the covariances and adjoint variables appearing in the Euler–Lagrange equations cannot be computed quickly. Fast convolution methods for standard covariances are given here; the methods are critical to the feasibility of data-subspace searches and hence generalized inversion itself. The section ends with some technical notes on rigorous inferences from penalty functionals, and on compounding covariances.

2.6.2 Least-squares smoothing of data; penalties for roughness

Let us set aside dynamics for now. Just consider interpolating some simple data $d_1, \ldots, d_M$, which are erroneous measurements of the scalar field $u = u(\mathbf{x})$ at the points $\mathbf{x}_1, \ldots, \mathbf{x}_M$. For simplicity, assume that $\mathbf{x}$ is planar: $\mathbf{x} = (x, y)$. We may define a quadratic penalty functional by

$$\mathcal{J}_0 = \mathcal{J}_0[u] = W_0 \iint_{\mathcal{D}} u^2 d\mathbf{x} + w|\mathbf{u} - \mathbf{d}|^2, \tag{2.6.1}$$

where $\mathcal{D}$ is some planar domain, W_0 and w are positive weights, and $\mathbf{u} = (u(\mathbf{x}_1), \ldots, u(\mathbf{x}_M))^T$. If $\hat{u} = \hat{u}(\mathbf{x})$ is an extremum of $\mathcal{J}$, then the calculus of variations implies that

$$W_0 \hat{u}(\mathbf{x}) = -w \boldsymbol{\delta}^T (\hat{\mathbf{u}} - \mathbf{d}), \tag{2.6.2}$$

where $\boldsymbol{\delta}^T = \boldsymbol{\delta}^T(\mathbf{x}) = (\delta(x - x_1)\delta(y - y_1), \ldots, \delta(x - x_M)\delta(y - y_M))$. So the "smallest" field that "nearly" fits the data is a crop of delta-functions. This is hardly useful. We would prefer a smoother field, so we should penalize the roughness of u, using

$$\mathcal{J}_1[u] = W_1 \iint_{\mathcal{D}} |\nabla u|^2 d\mathbf{x} + w|\mathbf{u} - \mathbf{d}|^2. \tag{2.6.3}$$

Extrema of $\mathcal{J}_1$ satisfy

$$W_1 \nabla^2 \hat{u} = w \boldsymbol{\delta}^T (\hat{\mathbf{u}} - \mathbf{d}). \tag{2.6.4}$$

So the field of least gradient which nearly fits the data is a crop of logarithms: recall that $\nabla^2 \ln |\mathbf{x}| = -(2\pi)^{-1}\delta(x)\delta(y)$. What's more, the solution of (2.6.4) is undefined up to harmonic functions ($\nabla^2 v = 0$) such as bilinear functions, which may or may not be fixed by boundary conditions. Logarithmic singularities are most likely undesirable, so we are led to consider

$$\mathcal{J}_2[u] = \iint \left[W_0 u^2 + W_1 |\nabla u|^2 + W_2 \left\{ \left(\frac{\partial^2 u}{\partial x^2}\right)^2 + 2\left(\frac{\partial^2 u}{\partial x \partial y}\right)^2 + \left(\frac{\partial^2 u}{\partial y^2}\right)^2 \right\} \right] d\mathbf{x},$$

$$+ w|\mathbf{u} - \mathbf{d}|^2, \tag{2.6.5}$$

which has extrema satisfying

$$W_2 \nabla^4 \hat{u} - W_1 \nabla^2 \hat{u} + W_0 \hat{u} = -w \boldsymbol{\delta}^T (\hat{\mathbf{u}} - \mathbf{d}). \tag{2.6.6}$$

Solutions of (2.6.6) behave like $|\mathbf{x} - \mathbf{x}_m|^2 \ln |\mathbf{x} - \mathbf{x}_m|$ for $\mathbf{x}$ near $\mathbf{x}_m$. This is usually acceptable. Evidently, any desired degree of smoothness may be achieved by imposing a sufficiently severe penalty for roughness. Note that the homogeneous equation corresponding to (2.6.6), subject to suitable boundary conditions, has only the trivial solution: $\hat{u} \equiv 0$.

2.6.3   Equivalent covariances

Now consider v, the Fourier transform of u:

$$v(\mathbf{k}) = \iint u(\mathbf{x})e^{i\mathbf{k}\cdot\mathbf{x}}\,d\mathbf{x}, \qquad (2.6.7a)$$

where the range of integration is the entire plane and $\mathbf{k} = (k, l)$. The inverse transform is

$$u(\mathbf{x}) = (2\pi)^{-2} \iint v(\mathbf{k})e^{-i\mathbf{k}\cdot\mathbf{x}}\,d\mathbf{k}. \qquad (2.6.7b)$$

The penalty functional (2.6.5) is equivalent to

$$\mathcal{J}_2[u] = (2\pi)^{-2} \iint (W_0 + W_1\,|\,\mathbf{k}\,|^2 + W_2\,|\,\mathbf{k}\,|^4)\,|\,v\,|^2\,d\mathbf{k} + w\,|\,\mathbf{u} - \mathbf{d}\,|^2, \qquad (2.6.8)$$

provided we assume that the domain $\mathcal{D}$ is the entire plane. Let the inverse transform of the reciprocal of the roughness weight in (2.6.8) be

$$C(\mathbf{x}) = (2\pi)^{-2} \iint (W_0 + W_1|\mathbf{k}|^2 + W_2|\mathbf{k}|^4)^{-1}e^{-i\mathbf{k}\cdot\mathbf{x}}\,d\mathbf{k}. \qquad (2.6.9)$$

After some calculus, it may be seen that (2.6.8) becomes

$$\mathcal{J}_2[u] = \iiiint u(\mathbf{x})W(\mathbf{x} - \mathbf{x}')u(\mathbf{x}')\,d\mathbf{x}\,d\mathbf{x}' + w\,|\,\mathbf{u} - \mathbf{d}\,|^2, \qquad (2.6.10)$$

where

$$\iint W(\mathbf{x} - \mathbf{x}')C(\mathbf{x}' - \mathbf{x}'')\,d\mathbf{x}' = \delta(\mathbf{x} - \mathbf{x}''). \qquad (2.6.11)$$

Thus there is close relationship between roughness penalties as in (2.6.5), and "nondiagonal sums" such as in (2.6.10). The latter penalty is in turn related to statistical estimation of a field having zero mean, and covariance

$$\overline{u(\mathbf{x})u(\mathbf{x}')} = C(\mathbf{x} - \mathbf{x}'). \qquad (2.6.12)$$

Exercise 2.6.1
Verify all the calculus sketched above, and show that C as defined in (2.6.9) only depends upon $|\mathbf{x}|$. That is, the random field u is *isotropic*.   □

Exercise 2.6.2
If $\mathcal{D}$ is bounded, what boundary conditions must $\hat{u}$ satisfy, in order to be an extremum of (2.6.5)?   □

Exercise 2.6.3 (Wahba and Wendelberger, 1980)
Express $\hat{u}$ in terms of representers. What is the associated inner product?   □

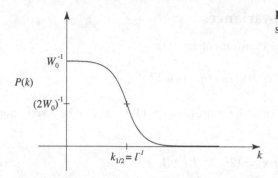

Figure 2.6.1 Power spectrum.

How should we choose W_0, W_1, and W_2? The inverse transform (2.6.9) yields, in particular, the hypothetical variance of $u(\mathbf{x})$:

$$\overline{u(\mathbf{x})^2} = C(0) = (2\pi)^{-2} \iint (W_0 + W_1 \mid \mathbf{k} \mid^2 + W_2 \mid \mathbf{k} \mid^4)^{-1} d\mathbf{k}, \qquad (2.6.13)$$

hence W_0 may be chosen to set the variance once W_1/W_0 and W_2/W_0 have been chosen. For example, let us assume that

$$W_1/W_0 = 0, \quad W_2/W_0 = l^4 \qquad (2.6.14)$$

for some length scale l. Then the hypothetical power spectrum of $u(\mathbf{x})$ is

$$P(k) = W_0^{-1}(1 + k^4 l^4)^{-1}, \qquad (2.6.15)$$

where $k = |\mathbf{k}|$. Defining the half-power point $k_{\frac{1}{2}}$ by

$$P(k_{\frac{1}{2}})/P(0) = \frac{1}{2} \qquad (2.6.16)$$

(see Fig. 2.6.1), we find that

$$k_{\frac{1}{2}} = l^{-1}. \qquad (2.6.17)$$

The functional (2.6.5), with parameters obeying (2.6.14), penalizes scales shorter than l ($k \gg k_{\frac{1}{2}} = l^{-1}$) and fits the data more closely if $W_0 \ll w$.

Exercise 2.6.4

The "bell-shaped" covariance

$$C(\mathbf{x}) = \exp(-\mid \mathbf{x} \mid^2 l^{-2}) \qquad (2.6.18)$$

is commonly used in optimal interpolation. Is there a corresponding smoothing norm, of the kind in (2.6.5)?   □

In summary, there are at least two ways of implementing least-squares smoothing: with covariances or with smoothing norms. These can be precisely or imprecisely matched, by choice of functional forms and parameters. It will be seen that the choice of implementation can be a matter of major convenience.

2.6.4   Embedding theorems

(The following two sections may be omitted from a first reading.) We began §2.6 with a discussion of quadratic penalty functionals used in the smoothing of data. It was seen that the smoothing field $\hat{u}(\mathbf{x})$ could have unacceptably singular behavior near the data points if the "smoothing norm" in the penalty functional were not chosen appropriately, that is, if the functional did not penalize derivatives of $u(\mathbf{x})$ of sufficiently high order. This was demonstrated by examining the solution of the Euler–Lagrange equation for $\hat{u}$, close to the data points. The examination was feasible since the functionals were quadratic in u and hence the Euler–Lagrange equations were linear, but we need not restrict ourselves in principle to quadratic functionals. There are powerful, theoretical guides that relate the mathematical smoothness of the estimate $\hat{u}$ to the differential order and algebraic power of the "smoothing norm" in the penalty functional. What follows is the crudest sketch of these so-called "embedding theorems" (Adams, 1975).

Let us suppose that the function $u = u(\mathbf{x})$ behaves algebraically near the point $\mathbf{x}_0$:

$$|u(\mathbf{x})| \sim K r^{\alpha} \tag{2.6.19}$$

for small r, where $r = |\mathbf{x} - \mathbf{x}_0|$, K is a positive constant and α is a positive or negative constant. The point $\mathbf{x}$ is in n-dimensional space: $\mathbf{x} \in R^n$. We unrigorously infer that any m^{th}-order partial derivative of u is also algebraic near $\mathbf{x}_0$, with

$$\left| D^{(m)} u(\mathbf{x}) \right| \sim K' r^{\alpha-m}, \tag{2.6.20}$$

where K' is another positive constant. Hence if we raise $D^{(m)}u$ to the power p and integrate over a bounded domain $\mathcal{D}$ that includes $\mathbf{x}_0$, then

$$\int_{\mathcal{D}} \cdots \int \left| D^{(m)} u(\mathbf{x}) \right|^p d\mathbf{x} \sim K'' \int_0^R r^{(\alpha-m)p+n-1} \, dr, \tag{2.6.21}$$

where R is the radius of $\mathcal{D}$. The integral on the rhs of (2.6.21) is finite, provided

$$(\alpha - m)p + n - 1 > -1. \tag{2.6.22}$$

That is, if the integral on the lhs of (2.6.21) is finite, then

$$|u(\mathbf{x})| \sim K r^{\alpha} < K r^{m-n/p} \tag{2.6.23}$$

for small r. Provided $mp > n > (m-1)p$, a rigorous treatment (Adams, 1975, p. 98) would replace the conclusion (2.6.23) with the more conservative inequality

$$|u(\mathbf{x}) - u(\mathbf{x}_0)| < K''' |\mathbf{x} - \mathbf{x}_0|^{\lambda}, \tag{2.6.24}$$

where

$$0 < \lambda \leq m - n/p. \tag{2.6.25}$$

If we were to include a term like the lhs of (2.6.21) in our smoothing norm, and were to find the $\hat{u}$ that minimizes the penalty functional, then we could conclude that the lhs of (2.6.21) would be finite, and hence (2.6.24) must hold. The positivity of λ in (2.6.25) ensures that u is at least continuous at x_0. If λ exceeds unity, then we can be sure that u is differentiable at x_0, and so on: $\lambda > k$ implies $D^{(k)}u$ is continuous at x_0.

For example, suppose $n = 2$ (we are in the plane: $x = (x, y)$); suppose $m = 2$ (we include second derivatives) and suppose $p = 2$ (we have a quadratic smoothing norm as in (2.6.5)); then

$$mp = 4 > n = 2 \not> (m - 1)p = 2, \tag{2.6.26}$$

and so we cannot even be sure that u is continuous. Nevertheless, we learned from the Euler–Lagrange equation (2.6.6) that $u \sim Kr^2 \ln r$, which is actually differentiable. Thus, the "embedding theorem" estimate of smoothness given in (2.6.25) is very conservative. The theorem would have us choose $p = 1.9$,

$$mp = 3.8 > n = 2 > (m - 1)p = 1.9. \tag{2.6.27}$$

Such a fractional power would make the calculus of variations very awkward, but the penalty functional would be well defined and would have a minimum $\hat{u}$ with guaranteed continuity.

2.6.5   Combining hypotheses: harmonic means of covariances

We have been considering penalty functionals, schematically of the form

$$\mathcal{J}[u] = (Mu) \circ C_f^{-1} \circ (Mu) + \cdots, \tag{2.6.28}$$

where C_f is the hypothesized covariance of Mu, M being some linear differential operator or linear model operator in general. We might also hypothesize that B_u is the covariance of u, in which case we could form the penalty functional

$$\mathcal{J}[u] = (Mu) \circ C_f^{-1} \circ (Mu) + u \circ B_f^{-1} \circ u + \cdots. \tag{2.6.29}$$

What now is the effectively hypothesized covariance for u? Manipulations like integrations by parts yield

$$(Mu) \circ C_f^{-1} \circ (Mu) = u \circ MC_f^{-1}M \circ u \tag{2.6.30}$$

$$= u \circ C_u^{-1} \circ u, \tag{2.6.31}$$

where

$$C_u = M^{-1}C_f M^{-1}. \tag{2.6.32}$$

Think of C_u as the covariance of solutions of the model $Mu = f$, where f has covariance C_f. We can now identify the effectively hypothesized covariance:

$$\mathcal{J}[u] = u \circ C_u^{-1} \circ u + u \circ B_u^{-1} \circ u + \cdots \qquad (2.6.33)$$

$$= u \circ A_u^{-1} \circ u, \qquad (2.6.34)$$

where

$$A_u = \left(C_u^{-1} + B_u^{-1}\right)^{-1} \qquad (2.6.35)$$

is the harmonic mean of the two covariances C_u and B_u.

Chapter 3

Implementation

It is a long road from deriving the formulae for the generalized inverse of a model and data to seeing results. First experiments (McIntosh and Bennett, 1984) involved a linear barotropic model separated in time, simple coarsely-resolved numerical approximations, a handful of pointwise measurements of sea level and a serial computer. Contemporary models of oceanic and atmospheric circulation involve nonlinear dynamics and parameterizations, advanced high-resolution numerical approximations, vast quantities of data often of a complex nature, and parallel computers. Chapter 3 introduces some general principles for travelling this long road of implementation.

The first principle is accelerating the representer algorithm by task decomposition, that is, by simultaneous computation of representers on parallel processors. The objective may be either the full representer matrix as required by the direct algorithm, or a partial matrix for preconditioning the indirect algorithm. The calculation of an individual representer, or indeed any backward or forward integration, may itself be accelerated by domain decomposition, but this is a common challenge in modern numerical computation (Chandra et al., 2001; Pacheco, 1996) and will not be addressed here. Even without considering the coarse grain of task decomposition or the fine grain of domain decomposition, the direct and indirect representer algorithms for linear inverses are highly intricate. Schematics are provided here in the form of "time charts".

Dynamical errors and input errors may be correlated in space or in time or in both. Error covariances must be convolved with adjoint variables. This is a massive task if four dimensions are involved and the numerical resolution is fine. Fast convolutions are critical to the scientific purpose of least-squares inversion, which is the testing of hypotheses about model errors. Posterior error statistics are equally essential, and are also massively expensive to compute and store in full detail. These statistics need not be computed with the same precision as the inverse itself, as they are only used for rough assessment of the likely accuracy of the inverse. Storage-efficient Monte Carlo algorithms permit computations of selected statistics with adequate reliability, on the same grid as the forward model if so desired.

Nonlinearity can only be overcome by iteration, but there is no unique way to iterate. This is a blessing in disguise, as certain choices for functional iterations can lead to linear, unbounded instability. No functional linearization yields statistical linearization, so significance tests and posterior error covariances that assume statistical linearity must be used with caution. Finally, crude parameterizations of unresolved natural processes may not be functionally smooth, thereby precluding variational assimilation. This obstacle should in principle be overcome by fiddling with the unnatural parameterization. Experience with trivial models suggests that we have much to learn.

3.1   Accelerating the representer calculation

3.1.1   So many representers . . .

The representer algorithm provides an explicit solution of linear Euler–Lagrange equations, and hence least-squares generalized inverses of overdetermined linear forward problems. There is one representer for each excess datum, and two model integrations are required (one backward, one forward) in order to construct each representer. (Note that we may regard the initial values and boundary values for the forward problem as data having exactly the same status as the finite set of measurements that overdetermine the forward problem; indeed, we may in principle envisage measurements obtained continuously along a track, and we shall in Chapter 6 consider specifying boundary values of too many components of a vector field.) It is impractical to compute every representer if their number is very large. There are rational approaches to reducing their number, as will be indicated in Chapter 5, but such approximations may not be necessary. It is possible to compute the representer solution for the inverse without reducing the number of representers, and without significant numerical approximation beyond that already implied by the numerical model. This technical advance has allowed the inversion of large data sets, with complex models imposed as weak constraints.

3.1.2   Open-loop maneuvering: a time chart

Recall again from §1.3.3 the representer solution for the inverse:

$$\hat{u}(x, t) = u_F(x, t) + \sum_{m=1}^{M} \hat{\beta}_m r_m(x, t), \tag{3.1.1}$$

where

$$(\mathbf{R} + \mathbf{C}_\epsilon)\hat{\beta} = \mathbf{h} \equiv \mathbf{d} - \mathcal{L}[u_F]. \tag{3.1.2}$$

Thus our tasks are:

(1) integrate the forward model for u_F ...   one integration;
(2) integrate the backward model for α ... M integrations;
(3) integrate the forward model for $\mathbf{r}$ ... M integrations;
for a total of   ... $I = 2M + 1$ integrations.

The backward and forward parts of the Euler–Lagrange equations are coupled by M numbers $\hat{u}_1, \ldots, \hat{u}_M$. The vector coefficient of the impulses in the adjoint equation (2.4.19), or coupling vector, is actually

$$\mathbf{C}_\epsilon^{-1}(\mathbf{d} - \mathcal{L}[\hat{u}]) = \mathbf{C}_\epsilon^{-1}\{\mathbf{d} - \mathcal{L}[u_F] - \mathbf{R}^{\mathsf{T}}\hat{\beta}\} \tag{3.1.3}$$
$$= \mathbf{C}_\epsilon^{-1}\{\mathbf{h} - \mathbf{R}^{\mathsf{T}}(\mathbf{R} + \mathbf{C}_\epsilon)^{-1}\mathbf{h}\}$$
$$\ldots \quad = (\mathbf{R} + \mathbf{C}_\epsilon)^{-1}\mathbf{h}$$
$$= \hat{\beta}. \tag{3.1.4}$$

That is, the coupling vector is the vector of representer coefficients. So we need not store the representer vector field $\mathbf{r}(x, t)$. We must compute $\mathbf{r}(x, t)$, measure it to obtain the representer matrix $\mathbf{R} = \mathcal{L}[\mathbf{r}^{\mathsf{T}}]$, solve (3.1.2) for $\hat{\beta}$, integrate the adjoint or backward EL equation (2.4.19) for $\hat{\lambda}(x, t)$ and then integrate the forward equation (1.5.14) for $\hat{u}(x, t)$. Now the integration count is $I = 2M + 3$. See Fig. 3.1.1 for a "time chart" implementing this so-called "open loop" version of the representer algorithm.

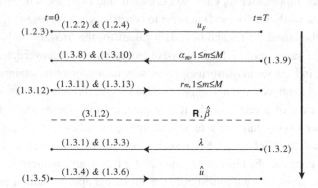

Figure 3.1.1 Time chart for implementing the representer algorithm with direct calculation of the representer coefficient $\hat{\beta}$, that is, by explicit or direct construction of the representer matrix $\mathbf{R}$. The heavy vertical arrow on the right indicates the order of execution, which starts at the top. Note that (3.1.1) need not be summed explicitly; once $\hat{\beta}$ is known, (3.1.4) resolves the coupling in (1.3.1)–(1.3.6). So in this "open loop" version, λ and hence $\hat{u}$ may be calculated with one backward integration and one forward integration. The representers r_m, $1 \leq m \leq M$, need not be stored. If the inverse weights W_f^{-1}, W_i^{-1} and w_b^{-1} are nondiagonal, then (1.3.11)–(1.3.13) and (1.3.4)–(1.3.6) require convolutions as in (1.5.14)–(1.5.16). See also (4.2.1)–(4.2.6) for a statement of the Euler–Lagrange equations for nondiagonal weighting.

Table 3.1.1 Processor work sheet.

Π_1	$\cdots$	Π_m	$\cdots$	Π_M
u_F		u_F		u_F
α_1		α_m		α_M
r_1		r_m		r_M
$\mathcal{L}[r_1]$		$\mathcal{L}[r_m]$		$\mathcal{L}[r_M]$
(all processors get the other $M-1$ columns of $\mathbf{R}$)				
(all processors solve for $\hat{\beta} = (\mathbf{R} + \mathbf{C}_\epsilon)^{-1}(\mathbf{d} - \mathcal{L}[u_F]))$				
λ		λ		λ
$\hat{u}$		$\hat{u}$		$\hat{u}$
$I = 5$		$I = 5$		$I = 5$

3.1.3 Task decomposition in parallel

The above task is ideally suited to parallel processing (Bennett and Baugh, 1992). Suppose we have M processors $\Pi_1, \ldots, \Pi_M$. The work sheet is as follows (see Table 3.1.1). The mth processor Π_m calculates the fields u_F, α_m and r_m, takes all M measurements $\mathcal{L}_1, \ldots, \mathcal{L}_M$ of r_m to obtain the mth column of $\mathbf{R}$, broadcasts this column to all of the other $M - 1$ processors, receives the other $M - 1$ columns in return, assembles $\mathbf{R}$, solves for the vector $\hat{\beta}$, then solves the Euler–Lagrange equations by calculating the field λ and the field $\hat{u}$.

So I is reduced from $2M + 3$ to 5 with an M-processor system. There is minimal exchange of data: each processor broadcasts one column of the representer matrix, and receives $M - 1$ columns in return. Each processor Π_m must have sufficient memory and speed for the computation and storage of $u(x, t)$, for $0 \le x \le L$ and $0 \le t \le T$. Note that the calculations of u_F, λ and $\hat{u}$ are M-fold redundant. This permits the programmer to release $M - 1$ processors during these steps; more importantly it permits a reduction of the number of broadcast messages by a factor of $M - 1$.

3.1.4 Indirect representer algorithm; an iterative time chart

Generalized inversion reduces exactly to solving the finite-dimensional system

$$(\mathbf{R} + \mathbf{C}_\epsilon)\hat{\beta} = (\mathbf{d} - \mathcal{L}[u_F]), \qquad (3.1.5)$$

or simply

$$\mathbf{P}\hat{\beta} = \mathbf{h}. \qquad (3.1.6)$$

A direct solution requires that $\mathbf{P}$ and hence $\mathbf{R}$ be explicitly known. However, the solution may be obtained iteratively, provided $\mathbf{P}\psi$ can be evaluated for any vector ψ. Then a standard iterative solver can convert a first-guess $\hat{\beta}_0$ into a solution $\hat{\beta} = \mathbf{P}^{-1}\mathbf{h}$.

Let us now examine how we could compute $\mathbf{P}\psi$, given any vector ψ. We have

$$\mathbf{P}\psi = \mathbf{R}\psi + \mathbf{C}_\epsilon\psi. \tag{3.1.7}$$

The data error covariance matrix $\mathbf{C}_\epsilon$ is explicitly known, so the nontrivial problem is the evaluation of $\mathbf{R}\psi$. The following procedure (Egbert *et al.*, 1994; Amodei, 1995; Courtier, 1997) does that *without* calculating the representers. First, solve the backward model, with coupling vector ψ:

$$-\frac{\partial\phi}{\partial t} - c\frac{\partial\phi}{\partial x} = \psi^{\mathrm{T}}\mathcal{L}[\delta\delta], \tag{3.1.8}$$

subject to

$$\phi = 0 \tag{3.1.9}$$

at $t = T$, and

$$\phi = 0 \tag{3.1.10}$$

at $x = L$. Second, solve the forward model, with adjoint field $\phi(x, t)$:

$$\frac{\partial\theta}{\partial t} + c\frac{\partial\theta}{\partial x} = C_f \bullet \phi, \tag{3.1.11}$$

subject to

$$\theta = C_i \circ \phi \tag{3.1.12}$$

at $t = 0$, and

$$\theta = cC_b * \phi \tag{3.1.13}$$

at $x = 0$. Comparison of (3.1.8)–(3.1.13), with the equations (2.4.14), (2.4.13) for α and $\mathbf{r}$ respectively, shows that

$$\phi(x, t) = \psi^{\mathrm{T}}\alpha(x, t), \quad \theta(x, t) = \psi^{\mathrm{T}}\mathbf{r}(\mathbf{x}, t) = \mathbf{r}(\mathbf{x}, t)^{\mathrm{T}}\psi. \tag{3.1.14}$$

Hence

$$\mathcal{L}[\theta] = \mathcal{L}[\mathbf{r}^{\mathrm{T}}]\psi = \mathbf{R}\psi, \tag{3.1.15}$$

which is just what is needed, at a cost of two integrations; see Fig. 3.1.2.

3.1.5   Preconditioners

If $\mathbf{P}$ is the unit matrix, then iterative solution of (3.1.6) should converge to $\mathbf{h}$ in one step. Hence iteration on (3.1.6) should in general be accelerated by premultiplying both sides of (3.1.6) with the inverse of a symmetric, positive-definite approximation to $\mathbf{P}$. That is, solve

$$\mathbf{P}_{\mathrm{A}}^{-1}\mathbf{P}\hat{\beta} = \mathbf{P}_{\mathrm{A}}^{-1}\mathbf{h}, \tag{3.1.16}$$

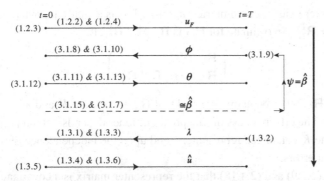

Figure 3.1.2 Time chart for implementing the indirect representer algorithm. The representer coefficients $\hat{\beta}$ are approximated by iterative solution of (3.1.2). Given a previous approximation ψ for $\hat{\beta}$, the "inner iteration" calculates $\mathbf{P}\psi = \mathbf{R}\psi + \mathbf{C}_\epsilon\psi$ with one backward integration and one forward integration (the representer matrix $\mathbf{R}$ is not explicitly constructed). This information is then used to find a better approximation ψ' for $\hat{\beta}$. Once $\hat{\beta}$ has been approximated with sufficient accuracy, λ and hence $\hat{u}$ are calculated as in the direct "open loop" algorithm. That is, the sum (3.1.1) is evaluated implicitly by one backward integration and one forward integration.

where $\mathbf{P}_A \cong \mathbf{P}$. Then (3.1.16) may be solved iteratively since, if we can evaluate $\mathbf{P}\psi$ for any ψ, then we can also evaluate $\mathbf{P}_A^{-1}\mathbf{P}\psi$. There are various choices for the pre-conditioner $\mathbf{P}_A$.

(i) (Bennett *et al.*, 1996) We could calculate all the representers quickly and cheaply on a coarse grid. Note that we would still be in effect solving (3.1.6) for the coefficients of the representers on the fine grid, so there would be no loss of resolution in $\hat{u}(x, t)$. However, the grid vertices may not coincide very closely with observing sites; the measurement functionals must involve interpolation formulae and these degrade appreciably as the grid gets coarser. That is, $\mathbf{P}_A$ may be a poor approximation to $\mathbf{P}$ and so convergence may not be greatly accelerated.

(ii) (Egbert and Bennett, 1996; Egbert, 1997) We could calculate some of the representers on the fine grid. Let $\mathbf{R}_C$ be the $M \times K$ matrix consisting of the first K columns of $\mathbf{R}$. That is, $\mathbf{R} = (\mathbf{R}_C, \mathbf{R}_{NC})$, where the non-calculated matrix $\mathbf{R}_{NC}$ is of dimension $M \times (M - K)$. Now $\mathbf{R}_C$ may be partitioned into an upper $K \times K$ block denoted by $\mathbf{R}_{11}$, and a lower $(M - K) \times K$ block $\mathbf{R}_{21}$, etc. That is,

$$\mathbf{R} = \begin{pmatrix} \mathbf{R}_{11} & \mathbf{R}_{12} \\ \mathbf{R}_{21} & \mathbf{R}_{22} \end{pmatrix} = (\mathbf{R}_C, \mathbf{R}_{NC}). \tag{3.1.17}$$

The representer matrix is symmetric, therefore $\mathbf{R}_{12}$ is actually known at this point: $\mathbf{R}_{12} = \mathbf{R}_{21}^{\mathsf{T}}$. An estimate for $\mathbf{R}_{22}$ is $\mathbf{R}_{21}\,\mathbf{R}_{11}^{-1}\,\mathbf{R}_{21}^{\mathsf{T}}$, thus

$$\mathbf{R}_A = \begin{pmatrix} \mathbf{R}_{11} & \mathbf{R}_{21}^{\mathsf{T}} \\ \mathbf{R}_{21} & \mathbf{R}_{21}\,\mathbf{R}_{11}^{-1}\,\mathbf{R}_{21}^{\mathsf{T}} \end{pmatrix}. \tag{3.1.18}$$

Then $\mathbf{P}_A = \mathbf{R}_A + \mathbf{C}_\epsilon$. Note that the ranks of $\mathbf{R}_A$ and $\mathbf{P}_A$ are K and M respectively. The effectiveness of this preconditioner depends upon a judicious choice for the K calculated representers, and upon the independence of the measurement errors.

(iii) Recall from (2.2.9) and (2.4.18) that the representer matrix is a covariance:

$$\mathbf{R} = \mathcal{L}C_v\mathcal{L}^{\mathsf{T}} \tag{3.1.19}$$

$$= E\{(\mathcal{L}[u] - \mathcal{L}[u_F])(\mathcal{L}[u] - \mathcal{L}[u_F])^{\mathsf{T}}\}. \tag{3.1.20}$$

Thus we may estimate $\mathbf{R}$ by Monte Carlo methods. That is, we make pseudo-random samples of $\mathcal{L}[u - u_F]$ and then evaluate sample covariances. The issue is: how many samples suffice?

Further details on implementation, including a flowchart, may be found in Chapter 5. The issue of sample size will be illustrated in §5.5.

3.1.6 Fast convolutions

We have seen the need to assume "nondiagonal" covariances for dynamical errors; that is,

$$C_f(\mathbf{x}, t, \mathbf{x}', t') \neq \delta(\mathbf{x} - \mathbf{x}')\delta(t - t'). \tag{3.1.21}$$

The covariance appears in the "forward" equation for the inverse estimate, for example (1.5.14):

$$\frac{\partial \hat{u}}{\partial t}(x, t) + c\frac{\partial \hat{u}}{\partial x}(x, t) = F(x, t) + (C_f \bullet \lambda)(x, t), \tag{3.1.22}$$

where

$$(C_f \bullet \lambda)(x, t) = \int_0^T ds \int_0^L dy\, C_f(x, t, y, s)\lambda(y, s). \tag{3.1.23}$$

Direct evaluation of this integral for each (x, t) would be prohibitively expensive in only one space dimension and time, and even more so in several space dimensions and time. Thus, a crucial requirement for smooth and hence physically acceptable inversions is an efficient algorithm for the evaluation of integrals such as (3.1.23). We shall refer to these loosely as "convolutions".

The following shortcut is very efficient (Derber and Rosati, 1989; Egbert et al., 1994). Assume that the covariance is purely spatial, and is "bell-shaped":

$$C(\mathbf{x}, \mathbf{x}') = C_0 \exp(-|\mathbf{x} - \mathbf{x}'|^2/L^2), \tag{3.1.24}$$

where C_0 is a constant. Assume that we wish to evaluate

$$b(\mathbf{x}) = \int\limits_{-\infty}^{\infty} \int\limits_{-\infty}^{\infty} C(\mathbf{x}, \mathbf{x}')a(\mathbf{x}') \, d\mathbf{x}'. \tag{3.1.25}$$

Solve the following pseudo-heat equation for $\theta = \theta(\mathbf{x}, s)$:

$$\frac{\partial \theta}{\partial s} = \nabla^2 \theta, \tag{3.1.26}$$

by time-stepping, subject to

$$\theta(\mathbf{x}, 0) = a(\mathbf{x}). \tag{3.1.27}$$

In two space dimensions, the solution is

$$\theta(\mathbf{x}, s) = (4\pi s)^{-1} \int\limits_{-\infty}^{\infty} \int\limits_{-\infty}^{\infty} \exp(-|\mathbf{x} - \mathbf{x}'|^2/(4s))a(\mathbf{x}') \, d\mathbf{x}'. \tag{3.1.28}$$

So let

$$s = L^2/4, \tag{3.1.29}$$

then

$$b(\mathbf{x}) = \pi L^2 C_0 \theta(\mathbf{x}, L^2/4). \tag{3.1.30}$$

Exercise 3.1.1
Compare the operation counts for numerical integration of (3.1.26), and numerical evaluation of (3.1.25), for one, two and three space dimensions. □

Exercise 3.1.2
How might you proceed when the spatial domain is finite? □

If the covariance is inhomogeneous, for example

$$C(\mathbf{x}, \mathbf{x}') = V(\mathbf{x})^{\frac{1}{2}} V(\mathbf{x}')^{\frac{1}{2}} \exp(-|\mathbf{x} - \mathbf{x}'|^2/L^2), \tag{3.1.31}$$

where $V(\mathbf{x}) = C(\mathbf{x}, \mathbf{x})$ is the variance, then proceed as above except that the initial condition becomes

$$\theta(\mathbf{x}, 0) = V(\mathbf{x})^{\frac{1}{2}} a(\mathbf{x}), \tag{3.1.32}$$

and the required result is

$$b(\mathbf{x}) = V(\mathbf{x})^{\frac{1}{2}} \pi L^2 \theta(\mathbf{x}, L^2/4). \tag{3.1.33}$$

Now consider temporal convolution, involving the simple form

$$C(t, t') = \exp(-|t - t'|/\tau).$$

That is, we wish to evaluate

$$b(t) = \int_0^T C(t, t') a(t') \, dt'. \tag{3.1.34}$$

This is the solution of

$$b_{tt} - \tau^{-2} b = -2\tau^{-1} a \tag{3.1.35}$$

for $0 \le t \le T$, subject to

$$b_t - \tau^{-1} b = 0 \tag{3.1.36}$$

at $t = 0$, and

$$b_t + \tau^{-1} b = 0 \tag{3.1.37}$$

at $t = T$. The two point boundary-value problem (3.1.35)–(3.1.37) is easily solved as two initial-value problems. First, solve

$$h_t + \tau^{-1} h = -2\tau^{-1} a \tag{3.1.38}$$

for $0 \le t \le T$, subject to

$$h = 0 \tag{3.1.39}$$

at $t = 0$. Then solve

$$b_t - \tau^{-1} b = h \tag{3.1.40}$$

for $0 \le t \le T$, subject to

$$b = -(\tau/2)h \tag{3.1.41}$$

at $t = T$.

Exercise 3.1.3
Show that the order of the two integrations in Exercise 3.1.2 may be reversed, with a modification to the terminal conditions.   □

3.2   Posterior errors

3.2.1   Strategy

How good is the generalized inverse $\hat{u}$? If u is the true circulation, and if we adopt the statistical interpretation of the inverse, then the error $u - \hat{u}$ has zero mean. There is a closed expression for the covariance of this error, or posterior error covariance. The expression involves the covariance of $u - u_F$ prescribed a priori in $\mathcal{H}_0$, and all of the representers. An efficient strategy for evaluating this formidable expression is essential.

The direct, serial representer algorithm requires the computation of M representers, one per datum. Each computation requires one backward and one forward integration; these may be executed in parallel if resources permit (see §3.1). It has been shown in §2.2 and §2.4 that the mth representer $r_m(x, t)$ is in fact the covariance of the mth measurement $\mathcal{L}_m[v]$ and the field $v(x, t)$ itself, where v is the response of the model to random forcing consistent with the hypothesis $\mathcal{H}_0$. The M representers having been computed, and stored, they may be used to construct error covariances for the inverse estimates $\hat{f}, \hat{i}, \hat{b}$ and $\mathcal{L}[\hat{u}]$ of the forcing, initial values, boundary values and measurements respectively (Bennett, 1992, §5.6). Computation of $\hat{u}$ using representers indirectly, as in §3.1.4, does not yield these posterior error covariances. The indirect approach typically requires about 10% of the effort of the direct approach; such efficiency is sometimes achieved by preliminary computation of the representers, either in part on the actual model grid or in the total on a coarser grid. This incomplete covariance information may suffice as an indication of the reliability of $\hat{u}$.

Regardless of the implementation of the representer algorithm, that is, either direct or indirect solution of the Euler–Lagrange equations for $\hat{u}$, it is possible to make "Monte Carlo" estimates of just as much covariance information as is required. The level of accuracy may be below that used to compute $\hat{u}$, but it is satisfactory as an indicator of the reliability of $\hat{u}$. The version of the Monte Carlo algorithm given in §3.2.5 is complicated, but it is highly memory-efficient.

3.2.2   Restatement of the "toy" inverse problem

For convenience, let us restate the "toy" problem here. The true ocean circulation u satisfies

$$\frac{\partial u}{\partial t}(x, t) + c\frac{\partial u}{\partial x}(x, t) = F(x, t) + f(x, t), \tag{3.2.1}$$

$$u(x, 0) = I(x) + i(x), \tag{3.2.2}$$

$$u(0, t) = B(t) + b(t), \tag{3.2.3}$$

where F, I and B are respectively the prior estimates of the forcing, initial values and boundary values (prior to assimilating data), while f, i and b are respectively the unknown errors in those priors. The prior estimate of u is u_F, which satisfies

$$\frac{\partial u_F}{\partial t}(x, t) + c\frac{\partial u_F}{\partial x}(x, t) = F(x, t), \tag{3.2.4}$$

$$u_F(x, 0) = I(x, t), \tag{3.2.5}$$

$$u_F(0, t) = B(t). \tag{3.2.6}$$

The data comprise an M-dimensional vector $\mathbf{d}$:

$$\mathbf{d} = \mathcal{L}[u] + \epsilon, \tag{3.2.7}$$

where $\mathcal{L}$ is a vector of linear measurement functionals and ϵ is the vector of the measurement errors. In order to improve upon u_F, we make an hypothesis $\mathcal{H}_0$ about the unknown errors f, i, b and ϵ:

$$Ef(x,t) = Ei(x) = Eb(t) = 0, \quad E\epsilon = \mathbf{0}; \tag{3.2.8}$$

$$\left.\begin{aligned} E(f(x,t)f(x',t')) &= C_f(x,t,x',t'), \\ E(i(x)i(x')) &= C_i(x,x'), \\ E(b(t)b(t')) &= C_b(t,t'), \\ E(\epsilon\epsilon^{\mathsf{T}}) &= \mathbf{C}_\epsilon, \end{aligned}\right\} \tag{3.2.9}$$

$$E(fb') = E(fi') = E(ib') = 0, \quad E(f\epsilon) = E(i\epsilon) = E(b\epsilon) = \mathbf{0}. \tag{3.2.10}$$

That is, we assume that the errors f, i, b and ϵ have vanishing means (F, I, B and $\mathbf{d}$ are unbiased) and have specified covariances C_f, C_i, C_b and $\mathbf{C}_\epsilon$. Then the posterior estimate $\hat{u}$ minimizes the estimator

$$\mathcal{J}[u] \equiv f \bullet C_f^{-1} \bullet f + i \circ C_i^{-1} \circ i + b * C_b^{-1} * b + \epsilon^{\mathsf{T}}\mathbf{C}_\epsilon^{-1}\epsilon, \tag{3.2.11}$$

where f, i, b and ϵ are related to u via (3.2.1)–(3.2.3) and (3.2.7). The symbols $\bullet$, $\circ$ and $*$ are defined by

$$\left.\begin{aligned} f \bullet g &\equiv \int_0^L dx \int_0^T dt\, f(x,t)g(x,t), \\ i \circ j &\equiv \int_0^L dx\, i(x)j(x), \\ a * b &\equiv \int_0^T dt\, a(t)b(t). \end{aligned}\right\} \tag{3.2.12}$$

The inverse covariances C_f^{-1}, C_i^{-1} and C_b^{-1} are defined in terms of (3.2.12):

$$\left.\begin{aligned} \int_0^L dx' \int_0^T dt'\, C_f^{-1}(x,t,x',t')C_f(x',t',x'',t'') &= \delta(x-x'')\delta(t-t''), \\ \int_0^L dx'\, C_i^{-1}(x,x')C_i(x',x'') &= \delta(x-x''), \\ \int_0^T dt'\, C_b^{-1}(t,t')C_b(t',t'') &= \delta(t-t''). \end{aligned}\right\} \tag{3.2.13}$$

The inverse of $\mathbf{C}_\epsilon$ is the standard matrix inverse $\mathbf{C}_\epsilon^{-1}$:

$$\mathbf{C}_\epsilon^{-1}\mathbf{C}_\epsilon = \mathbf{I}, \tag{3.2.14}$$

where $\mathbf{I}$ is the $M \times M$ unit matrix.

The minimizer of $\mathcal{J}$ and optimal estimate of u is

$$\hat{u}(x,t) = u_F(x,t) + \hat{\beta}^{\mathsf{T}}\mathbf{r}(x,t), \tag{3.2.15}$$

where the representer fields $\mathbf{r} = \mathbf{r}(x, t)$ and adjoint variables $\alpha = \alpha(x, t)$ satisfy

$$-\frac{\partial \alpha}{\partial t} - c\frac{\partial \alpha}{\partial x} = \mathcal{L}[\delta\delta], \tag{3.2.16}$$

$$\alpha = 0 \quad \text{at} \quad t = T, \tag{3.2.17}$$

$$\alpha = 0 \quad \text{at} \quad x = L, \tag{3.2.18}$$

$$\frac{\partial \mathbf{r}}{\partial t} + c\frac{\partial \mathbf{r}}{\partial x} = C_f \bullet \alpha, \tag{3.2.19}$$

$$\mathbf{r} = C_i \circ \alpha \quad \text{at} \quad t = 0, \tag{3.2.20}$$

$$\mathbf{r} = cC_b * \alpha \quad \text{at} \quad x = 0. \tag{3.2.21}$$

In (3.2.16), $\delta\delta = \delta(x - y)\delta(t - s)$ and $\mathcal{L}$ acts upon the (y, s) dependence. The representer coefficients $\hat{\beta}$ satisfy the linear system

$$\mathbf{P}\hat{\beta} \equiv (\mathbf{R} + \mathbf{C}_\epsilon)\hat{\beta} = \mathbf{d} - \mathcal{L}[u_F] \equiv \mathbf{h}, \tag{3.2.22}$$

where

$$\mathbf{R} = \mathcal{L}[\mathbf{r}^{\mathrm{T}}] = \mathcal{L}[\Gamma]\mathcal{L}^{\mathrm{T}},$$

$\Gamma = \Gamma(x, t, y, s)$ being the reproducing kernel ('rk'), or representer for a point measurement at (y, s).

3.2.3   Representers and posterior covariances

The error in the state estimate $\hat{u}$ is defined to be $u(x, t) - \hat{u}(x, t)$, and we would like to know its mean and covariance. First, let us note that the error in the prior estimate of the state has zero mean:

$$E(u(x, t) - u_F(x, t)) = 0, \tag{3.2.23}$$

and its covariance is

$$C_u(x, t, x', t') \equiv E((u(x, t) - u_F(x, t))(u(x', t') - u_F(x', t'))), \tag{3.2.24}$$

which is also, as was established in §2.2.3, the rk $\Gamma(x, t, x', t')$. The latter is, again, the representer for point measurement at (x', t'). Indeed (Exercise 2.2.1),

$$\Gamma = \gamma \bullet C_f \bullet \gamma + \gamma \circ C_i \circ \gamma + c^2\gamma * C_b * \gamma, \tag{3.2.25}$$

where $\gamma(x, t, x', t')$ is the influence function or Green's function for the "toy" model. (The symbol C_v in §2.2.3 has the same definition as C_u here in §3.2.3; it seems helpful to use different symbols in the two sections.) Notice that Γ, and hence C_u, is determined by the model, and by the prior covariances C_f, C_i and C_b. These are, again, the covariances of the errors f, i and b in the prior estimates F, I and B of the forcing, initial and boundary values, respectively. Recall also that

$$\mathbf{P} \equiv \mathbf{R} + \mathbf{C}_\epsilon = E(\mathbf{hh}^\mathsf{T}), \qquad (3.2.26)$$

where $\mathbf{h}$ is the prior data misfit:

$$\mathbf{h} = \mathbf{d} - \mathcal{L}[u_F]. \qquad (3.2.27)$$

Our main result is a tedious consequence of the above.

Exercise 3.2.1 (Bennett, 1992, §5.6; Xu and Daley, 2000)
Show that the state estimate is unbiased:

$$E(u(x, t) - \hat{u}(x, t)) = 0, \qquad (3.2.28)$$

and has as its covariance

$$C_{\hat{u}}(x, t, x', t') \equiv E((u(x, t) - \hat{u}(x, t))(u(x', t') - \hat{u}(x', t')))$$
$$= C_u(x, t, x', t') - \mathbf{r}^\mathsf{T}(x, t)\mathbf{P}^{-1}\mathbf{r}(x', t'). \qquad (3.2.29)$$

□

Recall that the optimal estimates of f, i, b and ϵ are

$$\hat{f}(x, t) = (C_f \bullet \lambda)(x, t), \qquad (3.2.30)$$

$$\hat{i}(x) = (C_i \circ \lambda)(x, 0), \qquad (3.2.31)$$

$$\hat{b}(t) = c(C_b * \lambda)(0, t) \qquad (3.2.32)$$

and

$$\hat{\epsilon} = \mathbf{d} - \mathcal{L}[\hat{u}], \qquad (3.2.33)$$

where $\lambda(x, t)$ is the weighted residual, or variable adjoint to $\hat{u}(x, t)$:

$$\lambda = C_f^{-1} \bullet \left(\frac{\partial \hat{u}}{\partial t} + c\frac{\partial \hat{u}}{\partial x} - F \right). \qquad (3.2.34)$$

It is readily shown that these all have zero mean:

$$E\hat{f} = E\hat{i} = E\hat{b} = 0, \quad E\hat{\epsilon} = 0. \qquad (3.2.35)$$

The posterior error covariances for f, i, b and ϵ follow easily from (3.2.29):

$$C_{\hat{f}}(x, t, x', t') \equiv E((f(x, t) - \hat{f}(x, t))(f(x', t') - \hat{f}(x', t')))$$
$$= C_f(x, t, x', t') - \mathbf{s}^\mathsf{T}(x, t)\mathbf{P}^{-1}\mathbf{s}(x', t'), \qquad (3.2.36)$$

where $\mathbf{s}$ is the representer residual vector:

$$\mathbf{s} \equiv \frac{\partial \mathbf{r}}{\partial t} + c\frac{\partial \mathbf{r}}{\partial x} ; \qquad (3.2.37)$$

$$C_{\hat{i}}(x, x') \equiv E((i(x) - \hat{i}(x))(i(x') - \hat{i}(x')))$$
$$= C_i(x, x') - \mathbf{r}^{\mathrm{T}}(x, 0)\mathbf{P}^{-1}\mathbf{r}(x', 0); \qquad (3.2.38)$$

$$C_{\hat{b}}(t, t') \equiv E((b(t) - \hat{b}(t))(b(t') - \hat{b}(t')))$$
$$= C_b(t, t') - \mathbf{r}^{\mathrm{T}}(0, t)\mathbf{P}^{-1}\mathbf{r}(0, t'), \qquad (3.2.39)$$

and

$$\mathbf{C}_{\hat{\epsilon}} \equiv E((\epsilon - \hat{\epsilon})(\epsilon - \hat{\epsilon})^{\mathrm{T}})$$
$$= \mathbf{R} - \mathbf{R}\,\mathbf{P}^{-1}\mathbf{R} \qquad (3.2.40)$$
$$= \mathbf{C}_{\epsilon} - \mathbf{C}_{\epsilon}\,\mathbf{P}^{-1}\mathbf{C}_{\epsilon}. \qquad (3.2.41)$$

Examination of (3.2.29)–(3.2.41) shows that, since C_f, C_i, C_b and $\mathbf{C}_{\epsilon}$ are prescribed, calculating the M representers $\mathbf{r}(x, t)$ yields $C_{\hat{f}}$, $C_{\hat{i}}$, $C_{\hat{b}}$ and $\mathbf{C}_{\hat{\epsilon}}$. Only $C_{\hat{u}}$ is not so available, since that requires $C_u = \Gamma$. We do know $C_{\hat{u}}$ at data sites: see (3.2.40). Thus the M representers give us the estimates $\hat{u}$, $\hat{f}$, $\hat{i}$, $\hat{b}$ and $\hat{\epsilon}$, and all the posterior error covariances except $C_{\hat{u}}$. The difference $C_u - C_{\hat{u}}$, or the "explained" covariance, may be expressed in terms of $\mathbf{r}$. In principle, we could calculate $C_u(x, t, x', t')$ as the rk $\Gamma(x, t, x', t')$, that is, by calculating the representer for every point (x', t'), but that is impractical. If the data were sufficiently dense, we could interpolate $\mathbf{C}_{\epsilon}$ to find $C_{\hat{u}}$ between data sites, but that is useful only if $\mathcal{L}$ involves just point measurement, and involves point measurement of every component when the state is multivariate (u, v, w, p, etc.).

3.2.4 Sample estimation

We may approximate the prior error covariance C_u using sample averages. Let the prior error be denoted by

$$v(x, t) \equiv u(x, t) - u_F(x, t). \qquad (3.2.42)$$

Then

$$\frac{\partial v}{\partial t} + c\frac{\partial v}{\partial x} = f, \qquad (3.2.43)$$
$$v = i \quad \text{at} \quad t = 0, \qquad (3.2.44)$$
$$v = b \quad \text{at} \quad x = 0. \qquad (3.2.45)$$

Use pseudo-random number generators to create pseudo-random fields $f(x, t)$, $i(x)$ and $b(t)$ consistent with the null hypothesis $\mathcal{H}_0$. For example, construct a "white-noise" field $w(x)$ satisfying

$$Ew(x) = 0, \quad E(w(x)w(x')) = \delta(x - x'), \qquad (3.2.46)$$

then "color" w to obtain a realization or sample for the initial error field i:

$$i(x) = \left(C_i^{\frac{1}{2}} \circ w\right)(x) = \int_0^L C_i^{\frac{1}{2}}(x, x') w(x')\, dx', \qquad (3.2.47)$$

where

$$C_i^{\frac{1}{2}} \circ C_i^{\frac{1}{2}} = C_i. \qquad (3.2.48)$$

Then

$$Ei(x) = 0, \quad E(i(x)i(x')) = C_i(x, x') \qquad (3.2.49)$$

as in $\mathcal{H}_0$. Samples of $f(x, t)$ and $b(t)$ may similarly be constructed.

In computational practice, the real variable x is replaced with a grid $x_n = n\Delta x$ for some uniform step Δx. Fast subroutines generate random numbers r independently and uniformly distributed in the interval $0 < r < 1$. Let $s = 2\sqrt{3}(r - \frac{1}{2})$; then $Es = 0$ and $E(s^2) = 1$. Let s_n be such a number; and let $w_n = s_n/\sqrt{\Delta x}$. Hence $Ew_n = 0$, and $E(w_n w_m) = \delta_{nm}/\Delta x \cong \delta(x_n - x_m)$. Then take $i_n = \sum_m C_{i_{nm}}^{\frac{1}{2}} w_m$. Generate K such samples of $i_n : i_n^1, i_n^2, \ldots, i_n^K$, and similarly generate f_{nl}^k, b_l^k, where l is a time index.

Approximate (3.2.43)–(3.2.45) on the (n, l) finite-difference grid. Integrate numerically to obtain samples v_{nl}^k for $k = 1, \ldots, K$. Then the sample prior error covariance is

$$C_u(x_n, t_l, x_p, t_q) \cong K^{-1} \sum_{k=1}^{K} v_{nl}^k v_{pq}^k. \qquad (3.2.50)$$

It is advisable to remove first any spurious sample mean of v_{nl}^k. Armed with this approximation to $C_u = \Gamma$, we may evaluate the representers $\mathbf{r} = \mathcal{L}[\Gamma]$ and hence *all* the posterior error covariances (3.2.29), (3.2.36)–(3.2.41).

Note 1. The prior data error covariance $\mathbf{C}_\epsilon$ influences the posteriors $\mathbf{C}_{\hat{u}}, \ldots, \mathbf{C}_{\hat{e}}$, but the actual data $\mathbf{d}$ do not.

Note 2. The posteriors $C_{\hat{u}}$, $C_{\hat{\imath}}$, $C_{\hat{b}}$, and $\mathbf{C}_{\hat{e}}$ given in (3.2.29), (3.2.38), (3.2.39) and (3.2.41) need not be related to a model; $C_{\hat{u}}$ is the posterior for the best linear unbiased estimate of u based on a prior u_F and data $\mathbf{d}$ having errors with zero means and covariances C_u and $\mathbf{C}_\epsilon$, respectively. Recall that $\mathbf{r} = \mathcal{L}[C_u]$. The measurement functionals $\mathcal{L}$ must be linear. The posterior (3.2.36) is valid only if f is related to u via a linear model.

Note 3. The posterior state estimate $\hat{u}$ may be expressed in terms of representers calculated as $\mathbf{r} = \mathcal{L}[C_u]$, where C_u is a sample covariance. However, a great many samples are needed for this approach to agree accurately with solutions of the representer equations (3.2.16)–(3.2.21) (Bennett *et al.*, 1998). The latter approach also requires many integrations, but the number of sample integrations should actually be compared to the cost of computing a preconditioner for an indirect representer solution as in §3.1.5.

Storage becomes a serious problem if C_u must be retained in full. It may suffice, for the purposes of indicating error levels, to compute C_u on a much coarser space–time grid than that used to calculate the state estimate $\hat{u}$.

3.2.5 Memory-efficient sampling algorithm

The following algorithm for sample estimates of $C_{\hat{u}}$ is memory-efficient but complicated:

   (i) generate samples of f^k, i^k and b^k, $k = 1, \ldots, K$;

  (ii) integrate to find samples for v^k, $k = 1, \ldots, K$;

 (iii) make a sample estimate of the representer matrix:

$$\mathbf{R} \cong \mathbf{R}_K = K^{-1} \sum_{k=1}^{K} \mathcal{L}[v^k]\mathcal{L}[v^k]^{\mathrm{T}}, \qquad (3.2.51)$$

     (note that it is not necessary to store all the K samples in order to evaluate (3.2.51), and note also that the rank of $\mathbf{R}_K$ is K);

 (iv) regenerate samples f^k, i^k and b^k, $k = 1, \ldots, K$, *identical* to those in (i);

  (v) recompute v^k, $k = 1, \ldots, K$;

 (vi) generate samples of measurement error ϵ^k, $k = 1, \ldots, K$;

(vii) derive sample data misfits:

$$\mathbf{h}^k = \mathcal{L}[v^k] + \epsilon^k \qquad (3.2.52)$$

     for $k = 1, \ldots, K$;

(viii) solve for the sample representer coefficients $\hat{\beta}^k$:

$$(\mathbf{R} + \mathbf{C}_\epsilon)\hat{\beta}^k = \mathbf{h}^k \qquad (3.2.53)$$

     for $k = 1, \ldots, K$ (use the indirect method of §3.1.4 and precondition with $(\mathbf{R}_K + \mathbf{C}_\epsilon)$);

 (ix) solve the Euler–Lagrange equations for the sample posterior state estimate

$$\hat{u}^k = u_F + (\hat{\beta}^k)^{\mathrm{T}}\mathbf{r} \qquad (3.2.54)$$

     for $k = 1, \ldots, K$;

  (x) evaluate the sample mean posterior error covariance:

$$C_{\hat{u}}(x, t, x', t') \cong K^{-1} \sum_{k=1}^{K} (u^k(x, t) - \hat{u}^k(x, t))(u^k(x', t') - \hat{u}^k(x', t'))$$

$$(3.2.55)$$

     (note that $u^k = u_F + v^k$).

This two-stage statistical simulation is intricate and requires many more model integrations than does single-stage simulation (3.2.50, etc.), but uses minimal memory without resorting to coarser grids.

Note 1. It is not necessary to evaluate (3.2.55) for all (x, t, x', t'); it may be evaluated as much as is needed, in order to indicate the reliability of $\hat{u}$.

Note 2. The posterior error means and covariances derive from the prior error moments assumed in $\mathcal{H}_0$ (see (3.2.8)–(3.2.10)). Rejection of $\mathcal{H}_0$ implies rejection of the posterior moments.

Exercise 3.2.2
Compare the computational requirements and storage requirements of the Monte Carlo algorithms given in §3.2.4 and in §3.2.5.   □

3.3   Nonlinear and nonsmooth estimation

3.3.1   Double, double, toil and trouble

Linear least-squares estimation problems may be solved efficiently by exploiting their linearity. The null subspace may be suppressed. Its complement, the data subspace, may be spanned with a finite basis – the representers. Their coefficients may be sought iteratively, and their sum may be formed without constructing each representer. Alas, nonlinearity is intrinsic to geophysical fluid dynamics, while many parameterizations involve functional nonsmoothness that precludes variational analysis. The general approach to smooth nonlinearity is to iterate yet again, leading to sequences of linear least-squares problems with solutions converging to that of the nonlinear least-squares problem. Several such "outer" iteration schemes are described here. The most orderly of them – the tangent linearization scheme – has the potential for grossly unphysical behavior. All linearization schemes are potentially unstable, drawing energy from the reference field and lacking amplitude modulation of that unstable growth. Practical experience of iterating on nonlinear Euler–Lagrange equations is not so bad: see Chapter 5. It seems that more data lead to faster convergence, while moderate smoothing of sources of linear instability in the adjoint equations can ensure stability at the price of slight suboptimality.

Nonlinearity in the dynamics vitiates the statistical analyses of §2.2 and §3.2. Dynamical linearization does not lead to statistical linearization, so the significance tests and recipes for posterior error covariances are suspect. In particular, bias can emerge in the inverse estimate of circulation even when none is present in the prior estimate of forcing.

Finally, nonsmoothness of the dynamics or the penalty functional precludes variational analysis. However, nonsmoothness is unnatural; it is an admission of poor resolution. Any mathematical "fudge" that removes it is entirely justified, since the nonsmoothness is itself a fudge.

What is really needed is not more theory, but more experience with realistic models and copious data. Nevertheless, here are some introductory analyses.

3.3.2   Nonlinear, smooth dynamics; least-squares

Consider a nonlinear wave equation:

$$\frac{\partial u}{\partial t} + \frac{\partial}{\partial x}\{U(u)u\} = F + f, \tag{3.3.1}$$

subject to an initial condition

$$u(x, 0) = I(x) + i(x) \tag{3.3.2}$$

at $t = 0$, and subject to the boundary condition

$$u(0, t) = B(t) + b(t) \tag{3.3.3}$$

at $x = 0$. As usual, F, I and B are priors, while f, i and b are unknown errors in the priors. The phase speed U is now a known function of the "ocean circulation" u. The form (3.3.1) is not the most general nonlinear wave equation, but it represents nondivergent advection in ocean models.

A simple penalty functional is

$$\mathcal{J}[u] = W_f \int_0^L dx \int_0^T dt\, f^2 + W_i \int_0^L dx\, i^2 + W_b \int_0^T dt\, b^2 + \cdots, \tag{3.3.4}$$

where the ellipsis denotes data penalties.

Exercise 3.3.1

Derive the following Euler–Lagrange equations for extrema of (3.3.4):

$$-\frac{\partial \lambda}{\partial t} - \hat{U}\frac{\partial \lambda}{\partial x} = \hat{u}\frac{d\hat{U}}{du}\frac{\partial \lambda}{\partial x} + (\cdots), \tag{3.3.5}$$

$$\lambda = 0 \tag{3.3.6}$$

at $t = T$,

$$\left[\hat{U} + \frac{d\hat{U}}{du}\hat{u}\right]\lambda = 0 \tag{3.3.7}$$

at $x = L$,

$$\frac{\partial \hat{u}}{\partial t} + \frac{\partial}{\partial x}\{\hat{U}\hat{u}\} = F + W_f^{-1}\lambda, \tag{3.3.8}$$

$$\hat{u} = I + W_i^{-1}\lambda \tag{3.3.9}$$

at $t = 0$, and

$$\hat{u} = B + W_b^{-1} \left[\hat{U} + \frac{d\hat{U}}{du} \hat{u} \right] \lambda \qquad (3.3.10)$$

at $x = 0$; $\hat{U} \equiv U(\hat{u})$. Note the "smoothness" assumption: U is differentiable with respect to u. In (3.3.5), $(\cdots)$ denotes data impulses. □

3.3.3 Iteration schemes

The system (3.3.5)–(3.3.10) is nonlinear, and so representers are of no immediate use. All kinds of iteration schemes suggest themselves, but the following two schemes have met with success:

Scheme A

$$-\frac{\partial \lambda_n}{\partial t} - \hat{U}_{n-1} \frac{\partial \lambda_n}{\partial x} = \hat{u}_{n-1} \left(\frac{d\hat{U}}{du} \right)_{n-1} \frac{\partial \lambda_{n-1}}{\partial x} + (\cdots)_n, \qquad (3.3.11)$$

$$\lambda_n = 0 \qquad (3.3.12)$$

at $t = T$,

$$\left[\hat{U}_{n-1} + \hat{u}_{n-1} \left(\frac{d\hat{U}}{du} \right)_{n-1} \right] \lambda_n = 0 \qquad (3.3.13)$$

at $x = L$,

$$\frac{\partial \hat{u}_n}{\partial t} + \frac{\partial}{\partial x} \{ \hat{U}_{n-1} \hat{u}_n \} = F + W_f^{-1} \lambda_n, \qquad (3.3.14)$$

$$\hat{u}_n = I + W_i^{-1} \lambda_n \qquad (3.3.15)$$

at $t = 0$, and

$$\hat{u}_n = B + W_b^{-1} \left[\hat{U}_{n-1} + \hat{u}_{n-1} \left(\frac{d\hat{U}}{du} \right)_{n-1} \right] \lambda_n \qquad (3.3.16)$$

at $x = 0$. The system (3.3.11)–(3.3.16) is linear in $\hat{u}_n$ and λ_n; it constitutes the Euler–Lagrange equations for a linear least-squares problem, and may be solved either with representers (Bennett and Thorburn, 1992) or with the sweep algorithm of §4.2.

Proving convergence of the sequence $\{\hat{u}_n, \lambda_n\}_{n=1}^{\infty}$ to a solution of (3.3.5)–(3.3.10) is most difficult in general, and has almost never been accomplished. Nevertheless, the sequence often seems to converge in practice, although the "source term" on the right-hand side of (3.3.11) may need spatial smoothing. If it is smoothed, $\hat{u}$ doesn't quite minimize $\mathcal{J}$. However, the approximate $\hat{u}$ suffices if $\hat{\mathcal{J}}$ is less than the expected value M, which is the number of data. The right-hand sides of (3.3.5) and (3.3.10) can cause difficulties because the calculus of the first variation involves a linearization of the dynamics, much as in a linear stability analysis. Specifically, the adjoint dynamics of (3.3.5) and its iterate (3.3.11) involve advective coupling, respectively, of λ and $\{\lambda_n\}_{n=1}^{\infty}$ to the reference flow, respectively $\hat{U}$ and $\{\hat{U}_{n-1}\}_{n=1}^{\infty}$, thus the adjoint dynamics can be destabilized. Note that (3.3.5) and (3.3.11) lack the potential for amplitude modulation that is present in the nonlinear forward dynamics (3.3.1).

Scheme B

$$-\frac{\partial \lambda_n}{\partial t} - \left[\hat{U}_{n-1} + \hat{u}_{n-1}\left(\frac{d\hat{U}}{du}\right)_{n-1}\right]\frac{\partial \lambda_n}{\partial x} = (\cdots)_n, \qquad (3.3.17)$$

$$\lambda_n = 0 \qquad (3.3.18)$$

at $t = T$,

$$\left[\hat{U}_{n-1} + \hat{u}_{n-1}\left(\frac{d\hat{U}}{du}\right)_{n-1}\right]\lambda_n = 0 \qquad (3.3.19)$$

at $x = L$,

$$\frac{\partial \hat{u}_n}{\partial t} + \frac{\partial}{\partial x}\left\{\hat{U}_{n-1}\hat{u}_n + \frac{d\hat{U}_{n-1}}{du}(\hat{u}_n - \hat{u}_{n-1})\hat{u}_{n-1}\right\} = F + W_f^{-1}\lambda_n, \qquad (3.3.20)$$

$$\hat{u}_n = I + W_i^{-1}\lambda_n \qquad (3.3.21)$$

at $t = 0$,

$$\hat{u}_n = B + W_b^{-1}\left[\hat{U}_{n-1} + \hat{u}_{n-1}\left(\frac{d\hat{U}}{du}\right)_{n-1}\right]\lambda_n \qquad (3.3.22)$$

at $x = 0$. This scheme, due to H.-E. Ngodock, employs the *tangent linearization* of (3.3.1) (Lions, 1971; Le Dimet and Talagrand, 1986). The linearized momentum equation (3.3.20) yields the Euler–Lagrange equation (3.3.17). The latter has no inhomogeneity on the rhs other than the usual impulses proportional to the data misfits of $\hat{u}_n$. The inhomogeneous term that appears on the rhs of (3.3.11) is now on the lhs of (3.3.17). That is, the term has become part of the adjoint operator. Furthermore, Scheme B would seem additionally risky, as it may introduce further linear instability into the forward dynamics (3.3.20). Scheme B obviates the need to compute and store a "first-guess" adjoint field λ_{F_n} that is the response to the "source term" in (3.3.11).

Note 1. There are many heuristic iteration schemes such as A. There is only one tangent linearization scheme B; it follows from the series expansion of the nonlinear flux in (3.3.1):

$$U(u_n)u_n = U(u_{n-1})u_{n-1} + \left\{\frac{dU}{du}(u_{n-1})\right\}(u_n - u_{n-1})u_{n-1}$$

$$+ U(u_{n-1})(u_n - u_{n-1}) + \cdots \qquad (3.3.23)$$

$$= \left\{\frac{dU}{du}\right\}_{n-1}(u_n - u_{n-1})u_{n-1} + U_{n-1}u_n + \cdots \qquad (3.3.24)$$

as appears in (3.3.20).

Note 2. If sequences of equations (3.3.11)–(3.3.16) or (3.3.17)–(3.3.22) are solved
using representers directly, then the latter must be recomputed for each iterate
(value of n). Alternatively, the iterative, indirect construction of the representer
solution must be repeated, for each such "linearizing" or "outer" iterate (value
of n). In principle, the indirect approach would require a recalculation of the
preconditioner for each outer iterate, but in practice such effort does not seem
necessary for $n > 2$.

Note 3. Since U is a smooth function of u, we can calculate the gradient of $\mathcal{J}$ with
respect to $u(x, t)$; for example,

$$\frac{\delta \mathcal{J}}{\delta u(x, t)} = -2 \left\{ \frac{\partial \lambda}{\partial t} + \left(\frac{dU}{du} u + U \right) \frac{\partial \lambda}{\partial x} \right\}, \qquad (3.3.25)$$

where

$$\lambda \equiv W_f \left(\frac{\partial u}{\partial t} + \frac{\partial}{\partial x} \{Uu\} - F \right), \qquad (3.3.26)$$

if $0 < x < L, 0 < t < T$ and (x, t) is not a data point. (Readers unfamiliar with
functional differentiation as in (3.3.25) may prefer to revisit this section after
studying the discrete analog in §4.1.) Thus, given the field $u = u(x, t)$ we can
evaluate $\lambda(x, t)$ and hence the gradient of $\mathcal{J}$, enabling a gradient search for the
field $\hat{u} = \hat{u}(x, t)$ that satisfies

$$\frac{\delta \mathcal{J}}{\delta u}[\hat{u}] = 0. \qquad (3.3.27)$$

Only one level of iteration is needed for this "state space" search: there is no
need for two levels as in the doubly iterated representer approach or "data space
search". However, preconditioning is still essential for a state space search; in
effect the inverse of the Hessian form

$$H \equiv \frac{\delta^2 \mathcal{J}}{\delta u(x, t) \delta u(y, s)} \qquad (3.3.28)$$

is required. Calculating H in full is usually prohibitive, as is inverting H. Some
approximations, such as replacing H with its diagonal, do seem useful. See also
§4.1.5.

3.3.4   Real dynamics: pitfalls of iterating

The idealized nonlinear wave dynamics of (3.3.1) provide a conveniently simple setting
for the introduction of iterative solution schemes. The linear dynamics of Scheme A,
as displayed in (3.3.14), retain the character of those in (3.3.1). However, the linear
dynamics of Scheme B as shown in (3.3.20) are, as already indicated, of a different
character. This can have radical consequences for real dynamics.

(i) *Continuity*

Consider first an equation for conservation of volume, as appears in shallow-water models, layered models (Bleck and Smith, 1990) or indeed any reduced-gravity Primitive Equation model (e.g., Gent and Cane, 1989):

$$\frac{\partial h}{\partial t} + \mathbf{u} \cdot \nabla h + h \nabla \cdot \mathbf{u} = 0, \tag{3.3.29}$$

where $\mathbf{x} = (x, y)$, $\nabla = (\frac{\partial}{\partial x}, \frac{\partial}{\partial y})$, $\mathbf{u} = (u, v)$, $h = h(\mathbf{x}, t)$ and $\mathbf{u} = \mathbf{u}(\mathbf{x}, t)$. Defining $\mathbf{X}(\mathbf{a}|t)$ to be the position at time t of a fluid particle that was initially at position $\mathbf{a}$, that is,

$$\frac{d\mathbf{X}}{dt}(\mathbf{a}|t) = \mathbf{u}(\mathbf{X}, t), \tag{3.3.30}$$

subject to

$$\mathbf{X}(\mathbf{a}|0) = \mathbf{a}, \tag{3.3.31}$$

and defining $h(\mathbf{a}|t)$ and $\mathbf{u}(\mathbf{a}|t)$ by

$$h(\mathbf{a}|t) \equiv h(\mathbf{X}(\mathbf{a}|t), t), \quad \mathbf{u}(\mathbf{a}|t) \equiv \mathbf{u}(\mathbf{X}(\mathbf{a}|t), t) \tag{3.3.32}$$

allows us to express (3.3.29) as

$$\frac{Dh}{Dt}(\mathbf{a}|t) + h(\mathbf{a}|t)(\nabla \cdot \mathbf{u})(\mathbf{a}|t) = 0, \tag{3.3.33}$$

where the Lagrangian derivative is

$$\frac{Dh}{Dt}(\mathbf{a}|t) \equiv \left\{ \frac{\partial h}{\partial t}(\mathbf{x}, t) + \mathbf{u}(\mathbf{x}, t) \cdot \nabla h(\mathbf{x}, t) \right\}_{\mathbf{x}=\mathbf{X}(\mathbf{a}|t)}. \tag{3.3.34}$$

The formal solution of (3.3.33) is

$$h(\mathbf{a}|t) = h(\mathbf{a}|0) \exp \left\{ -\int_0^t (\nabla \cdot \mathbf{u})(\mathbf{a}|s) \, ds \right\}, \tag{3.3.35}$$

which, so long as $\nabla \cdot \mathbf{u}$ remains integrable in time, cannot change sign. The ocean cannot "dry out", nor can ocean layers "outcrop" in a finite time. With small-amplitude gravity waves in mind, it is tempting to apply Scheme A to (3.3.29) as follows:

$$\frac{\partial h_n}{\partial t} + \mathbf{u}_{n-1} \cdot \nabla h_n = -h_{n-1} \nabla \cdot \mathbf{u}_n. \tag{3.3.36}$$

Together with a matching linearization of the momentum equations, (3.3.36) would capture such waves. However, the relegation of the divergence to a source term, as far as (3.3.36) alone is concerned, may cause h_n to change sign just as though it were the perturbation amplitude of a small wave. Alternatively, applying Scheme A as follows:

$$\frac{\partial h_n}{\partial t} + \mathbf{u}_{n-1} \cdot \nabla h_n + h_n \nabla \cdot \mathbf{u}_{n-1} = 0, \tag{3.3.37}$$

preserves the positivity of h_n. Scheme B leads uniquely to

$$\frac{\partial h_n}{\partial t} + \mathbf{u}_{n-1} \cdot \nabla h_n + h_n \nabla \cdot \mathbf{u}_{n-1}$$
$$= -(\mathbf{u}_n - \mathbf{u}_{n-1}) \cdot \nabla h_{n-1} - h_{n-1} \nabla \cdot (\mathbf{u}_n - \mathbf{u}_{n-1}), \tag{3.3.38}$$

which does not ensure positivity of h_n.

(ii) *Thermodynamics*

Consider the turbulent transfer of heat, modeled simply by

$$\frac{\partial T}{\partial t} = \frac{\partial}{\partial z}\left(K \frac{\partial T}{\partial z}\right), \tag{3.3.39}$$

where the positive eddy conductivity K is a function of the temperature gradient:

$$K = K\left(\frac{\partial T}{\partial z}\right) > 0. \tag{3.3.40}$$

It is commonly assumed that K is a function of the gradient Richardson number (see, for example, Pacanowski and Philander, 1981), but it suffices for this discussion to consider just (3.3.40). Linearizing (3.3.39) with Scheme A leads naturally to

$$\frac{\partial T_n}{\partial t} = \frac{\partial}{\partial z}\left(K_{n-1}\frac{\partial T_n}{\partial z}\right), \tag{3.3.41}$$

where $K_{n-1} \equiv K(\partial T_{n-1}/\partial z) > 0$. Both (3.3.39) and (3.3.41) yield well-posed initial-value problems for $t > 0$. Scheme B leads uniquely to

$$\frac{\partial T_n}{\partial t} = \frac{\partial}{\partial z}\left\{\left(K_{n-1} + K'_{n-1}\frac{\partial T_{n-1}}{\partial z}\right)\frac{\partial T_n}{\partial z}\right\} - K'_{n-1}\left(\frac{\partial T_{n-1}}{\partial z}\right)^2, \tag{3.3.42}$$

where $K'(\theta) = dK(\theta)/d\theta$, which is commonly assumed to be negative, for $\theta > 0$. Indeed, it is possible for $(K_{n-1} + K'_{n-1}\partial T_{n-1}/\partial z)$ to change sign in the tropical Pacific Ocean, rendering (3.3.42) ill-posed for forward integration. The associated Euler–Lagrange equation would therefore be ill-posed for backward integration.

The preceding examples show that while tangent linearization ("Scheme B") has the merits of unique definition and efficient implementation, it can in principle lead to unrealistic dynamics. If variational assimilation with a realistic model leads to difficulties, it is advisable to experiment with the linearization scheme.

3.3.5 Dynamical linearization is not statistical linearization

Consider again the nonlinear wave equation (3.3.1), subject to some initial and boundary conditions that need not be considered here explicitly. The prior solution u_F obeys

$$\frac{\partial u_F}{\partial t} + \frac{\partial}{\partial x}\{U(u_F)u_F\} = F. \tag{3.3.43}$$

Regarding the solution u of (3.3.1) as the true circulation, the error in u_F is $v = u - u_F$. It follows that

$$\frac{\partial v}{\partial t} + \frac{\partial}{\partial x}\{U(u_F + v)(u_F + v) - U(u_F)u_F\} = f. \tag{3.3.44}$$

It has been assumed in the analyses of preceding chapters that $Ef = 0$, that is, F is an unbiased estimate of the true forcing $F + f$. (We may always assume that the hypothetical field Ef vanishes, as a nonvanishing field may be absorbed into F. The hypothesized mean, vanishing or nonvanishing, may of course be wrong.) If U were constant, then (3.3.44) would become

$$\frac{\partial v}{\partial t} + U\frac{\partial v}{\partial x} = f. \tag{3.3.45}$$

Hence, as the expectation E is a linear operator:

$$\frac{\partial(Ev)}{\partial t} + U\frac{\partial(Ev)}{\partial x} = 0, \tag{3.3.46}$$

for which the solution is $Ev = 0$, subject to suitable (linear, unbiased) initial and boundary conditions. In the general case U depends upon u, and f is not identically zero, thus it cannot be concluded that $Ev = 0$. The statistical variability of the forcing f can induce a bias in the circulation u, even though the prior estimate of forcing is unbiased. Moreover, closed forms such as (2.2.8) are no longer available for the covariance of u.

Now consider a simple iteration of (3.3.1) about the iterate $\hat{u}_{n-1}$ for an inverse estimate $\hat{u}$:

$$\frac{\partial u}{\partial t} + \frac{\partial}{\partial x}\{U(\hat{u}_{n-1})u\} = F + f. \tag{3.3.47}$$

The prior solution u_{F_n} satisfies

$$\frac{\partial u_{F_n}}{\partial t} + \frac{\partial}{\partial x}\{U(\hat{u}_{n-1})u_{F_n}\} = F, \tag{3.3.48}$$

hence the prior error $v_n = u - u_{F_n}$ satisfies

$$\frac{\partial v_n}{\partial t} + \frac{\partial}{\partial x}\{U(\hat{u}_{n-1})v_n\} = f. \tag{3.3.49}$$

The operator in (3.3.49) is dynamically linear. A linear superposition of forcing yields a linear superposition of solutions. That is the case, as long as the dependence of $U(\hat{u}_{n-1})$ upon the actual forcing error f is ignored. In fact $\hat{u}_{n-1}$ is a linear combination of representers with coefficients proportional to the prior data misfit. The latter is of the form

$$\mathbf{d} - \mathcal{L}[u_{F_{n-1}}], \tag{3.3.50}$$

where $\mathbf{d}$ is the data. Of course, $u_{F_{n-1}}$ is dependent on f via $\hat{u}_{n-2}$, and so on, but it suffices here to note just that

$$\mathbf{d} = \mathcal{L}[u] + \epsilon, \tag{3.3.51}$$

where u satisfies (3.3.1) and ϵ is the vector of measurement errors. The forcing error f that appears in the nonlinear model (3.3.1) is the same as the forcing error f that appears in the iterated error model (3.3.49). In particular,

$$E\{U(\hat{u}_{n-1})v_n\} \neq U(\hat{u}_{n-1})Ev_n, \tag{3.3.52}$$

and so we cannot conclude that Ev_n vanishes identically when Ef does. The dynamically linear model (3.3.49) is statistically nonlinear. Significance tests and posterior error covariances for inverses of (3.3.47), without regard to its statistical nonlinearity, are at best guides rather than rigorous results. A strong warning from these guides, to the effect that the hypothesized prior means and covariances for f and for the initial and boundary inputs are unreliable, should nevertheless be taken seriously.

3.3.6   Linear, smooth dynamics; non-least-squares

Suppose that the model is our original linear wave equation (1.2.6) and ancillary information (1.2.7)–(1.2.8), but suppose that our estimator or penalty functional is, for whatever reason, quartic in the residuals:

$$Q[u] = K_f \int_0^T dt \int_0^L dx\, f^4 + K_b \int_0^T dt\, b^4 + K_i \int_0^L dx\, i^4 + k \sum_{m=1}^M \epsilon_m^4. \tag{3.3.53}$$

The gradient of Q with respect to $u(x, t)$ is still well-defined; for example,

$$\frac{\delta Q}{\delta u(x, t)} = -4\left\{ \frac{\partial \mu}{\partial t} + c\frac{\partial \mu}{\partial x} \right\}, \tag{3.3.54}$$

where

$$\mu = K_f \left(\frac{\partial u}{\partial t} + c\frac{\partial u}{\partial x} - F \right)^3, \tag{3.3.55}$$

if $0 < x < L$, $0 < t < T$ and (x, t) is not a data point. Of course, we could express (3.3.54) in terms of $\mu^{\frac{1}{3}}$. A gradient-search in state space is in principle indifferent to the nonlinearity of (3.3.53) and (3.3.54), but the nonlinearity of the Euler–Lagrange equations for nonquadratic penalties precludes the use of representers without a linearizing iteration. Attempts to do so have not been reported in a meteorological or oceanographic context.

Exercise 3.3.2

Derive the Euler–Lagrange equations for extrema of Q defined by (3.3.53).   □

3.3.7   Nonsmooth dynamics, smooth estimator

Suppose that the phase velocity in our first-order wave equation is a nonsmooth function of the circulation, for example

$$\frac{\partial u}{\partial t} + \frac{\partial}{\partial x}\{U(u)u\} = F + f, \tag{3.3.56}$$

where

$$U(u) = \begin{cases} u & u > 0 \\ 0 & u < 0. \end{cases} \tag{3.3.57}$$

This type of continuous but nondifferentiable parameterization of convection is especially common in mixed-layer models, and in models of convective adjustment. See for example Zebiak and Cane (1987) and Cox and Bryan (1984), respectively. The circulation variable u is usually temperature, while the "mixing-function" U depends upon the vertical velocity, which is related to temperature via the dynamics.

The difficulty is that the first variation of f with respect to u is not defined at $u = 0$. The situation may be handled using engineering "optimal control theory", but that seems misguided in this context. There really are nonsmooth dependencies in engineering, but in geophysical fluid dynamics we are merely making a crude parameterization of unresolved processes. Consider a finite-difference approximation to (3.3.56). The values of u at grid points are representatives of values within intervals. Yet, in reality, there will be a range of values within an interval, and convection will not start everywhere simultaneously. It therefore seems more sensible to replace a nonsmooth dependence as in (3.3.57) with a smooth dependence such as

$$U(u) = \epsilon \ln (1 + e^{u/\epsilon}), \tag{3.3.58}$$

where ϵ is small. Clearly,

$$\left. \begin{aligned} &U \sim u && \text{as } u \to \infty, \\ &U = \epsilon \ln 2 && \text{at } u = 0, \\ &\text{and} \\ &U \sim \epsilon e^{u/\epsilon} \sim 0 && \text{as } u \to -\infty. \end{aligned} \right\} \tag{3.3.59}$$

Then

$$\frac{dU}{du} = \frac{e^{u/\epsilon}}{1 + e^{u/\epsilon}}, \tag{3.3.60}$$

so

$$
\left.\begin{aligned}
\frac{dU}{du} &\sim 1 && \text{as } u \to \infty, \\
\frac{dU}{du} &= \tfrac{1}{2} && \text{at } u = 0, \\
\text{and} \\
\frac{dU}{du} &\sim e^{u/\epsilon} \sim 0 && \text{as } u \to -\infty.
\end{aligned}\right\} \tag{3.3.61}
$$

In particular, $\frac{dU}{du} < 1$ for all u. Had there been a discontinuity in U at $u = 0$, say

$$
U = \begin{cases} c & u > 0 \\ 0 & u < 0, \end{cases} \tag{3.3.62}
$$

then, while it would be possible to smooth over the discontinuity in some small interval around $u = 0$, the derivative $\frac{dU}{du}$ would be very large in that interval. In any case (3.3.62) would seem an implausible parameterization of a gfd process.

Even the mildest departure from nonsmoothness can greatly complicate an otherwise simple variational problem. For a lengthy investigation, see Xu and Gao (1999) and the references therein. Miller, Zaron, and Bennett (1994) consider varying the times of onset and end of convective mixing in a trivial linear model. The resulting variational problem is highly nonlinear, and the solution is not unique unless there are penalties for errors in prior estimates of the timing of convection. More recently, Zhang *et al.* (2000) find that imposing smoothness on even a trivial model by "fiddling" may do more computational harm than good. They also explore the concept of subgradients of nonsmooth functionals, and examine the resulting generalizations of gradient searches in state space.

3.3.8   Nonsmooth estimator

Now suppose that the estimator for (3.3.56) is not smooth, such as:

$$
\mathcal{J}[u] = \int_0^T dt \int_0^L dx \, |f| + \cdots, \tag{3.3.63}
$$

or

$$
\mathcal{J}[u] = \int_0^T dt \int_0^L dx \, \text{sgn}\{f_0 - |f|\}, \tag{3.3.64}
$$

where

$$
\text{sgn}(x) = \begin{cases} 1 & x > 0 \\ -1 & x < 0. \end{cases} \tag{3.3.65}
$$

Neither of these functionals has uniformly well-defined first variations with respect to u. Consequently there is no gradient information that can guide a search, and Euler–Lagrange conditions cannot be formulated, unless we are sure of special information such as the sign of $\hat{f}$.

Only brute-force minimization is available in the absence of gradient information, but brute force can be applied thoughtfully: see the method of simulated annealing in §4.4.

Chapter 4

The varieties of linear and nonlinear estimation

A data space search is the most efficient way to solve a linear, least-squares smoothing problem defined over a fixed time interval. The method exploits linearity, and so is unavailable for nonlinear dynamics, or for penalties other than least-squares. As discussed in Chapter 3, a data space search may be conducted on linear iterates of the nonlinear Euler–Lagrange equations. The existence of the nonlinear equations implies that the penalty is a smooth functional of the state, in which case a state space search may always be initiated. The nature of state space searches is intuitively clear, and their use is widespread. Conditioning degrades as the size of the state space gets very large. Collapsing the size of the state space by assuming "perfect" dynamics is the basis of "the" variational adjoint method: only initial values, boundary values and parameter values are varied. Preconditioning may in principle be effected by use of second-order variational equations, but even iterative construction of the state space preconditioner is unfeasible for highly realistic problems. Technique for numerical integration of variational equations is not a paramount consideration, but deserving of attention since it can be consuming of human time.

Operational forecasting is inherently sequential; data are constantly arriving and forecasts must be issued regularly. In such an environment, it is more natural to filter a model and data sequentially than to smooth them over a fixed interval. The Kalman filter is available for linear dynamics; it forecasts the covariance of the error in the forecast of the state, and uses the covariance to make a least-squares spatial interpolation of newly arriving data. If the dynamics are nonlinear, the Kalman filter may be applied iteratively. It is computationally expensive, even in the linear case, and various economies are practised.

Least-squares is the estimator of maximum likelihood for Gaussian or normal random quantities. It may be applied to the estimation of any quantity, but the reliability of the estimate would be in doubt. Maximum likelihood estimators are defined for any random variable, yet are subtle or even ambiguous for even the simplest of non-normal distributions. Monte Carlo methods for simulating random quantities have great appeal, owing to their conceptual simplicity. Ingenious algorithms exist for generating random variables of any distribution. The same algorithms lead to equally ingenious optimization methods for arbitrary penalties.

4.1 State space searches

4.1.1 Gradients

The representer algorithm described in the previous chapters is an arcane and intricate method for minimizing quadratic penalty functionals such as $\mathcal{J}$ in (1.2.9). It would be much simpler to evaluate the gradient of $\mathcal{J}$ with respect to changes in the errors f, i and b, and then use the gradient information in a search for the minimum of $\mathcal{J}$. Naïve evaluation of the gradient would be prohibitively expensive, but an economical alternative exists.

We may avoid the abstraction of a gradient in function space, by referring to numerical models. Then the gradient becomes a finite-dimensional vector with as many components as there are independently variable quantities, or "controls". These may be as numerous as the entire set of gridded variables, in which case the search would be hopelessly ill-conditioned. Imposing the dynamics as a strong constraint reduces the set of controls to the initial conditions, boundary conditions and dynamical parameters such as diffusivities. Even so, these may be numerous and convergence may be slow. Crude preconditioners are unreliable, while full preconditioning remains unfeasible.

The section concludes with some hints on deriving gradients.

4.1.2 Discrete penalty functional for a finite difference model; the gradient

For simplicity, the following discussion is based on an "ocean model" of maximal simplicity. It is the ordinary differential equation

$$\frac{du}{dt}(t) = F(t) + f(t), \tag{4.1.1}$$

where $0 \leq t \leq T$, subject to

$$u(0) = I + i. \tag{4.1.2}$$

In (4.1.1) and (4.1.2), F and I denote prior estimates of forcing and initial values, while f and i denote errors in those estimates. It suffices to consider a single datum:

$$u(t_d) = U + \epsilon, \tag{4.1.3}$$

where U is the datum at time t_d, and ϵ is the measurement error. A least-squares estimator for f, i, ϵ and hence u is

$$J[u] = W_f \int_0^T dt \ f^2 + W_i i^2 + w\epsilon^2 \tag{4.1.4}$$

$$= W_f \int_0^T dt \left(\frac{du}{dt} - F\right)^2 + W_i(u(0) - I)^2 + w(u(t_d) - U)^2. \tag{4.1.5}$$

We may avoid functional analysis by replacing the integrals and derivatives in (4.1.5) with sums and differences. To this end, define

$$t_n = n\Delta t \tag{4.1.6}$$

for $0 \le n \le N$, where

$$\Delta t = T/N; \tag{4.1.7}$$

hence

$$t_0 = 0, \quad t_N = T. \tag{4.1.8}$$

Assume

$$t_d = n_d \Delta t \tag{4.1.9}$$

for some integer n_d:

$$0 < n_d < N. \tag{4.1.10}$$

Define

$$u_n \equiv u(t_n) \tag{4.1.11}$$

for $0 \le n \le N$, and approximate derivatives with forward differences:

$$\left(\frac{du}{dt}\right)_n \simeq \frac{u_{n+1} - u_n}{\Delta t}. \tag{4.1.12}$$

Replacing the integral with a sum, (4.1.5) becomes

$$J_N = W_f(\Delta t)^{-1} \sum_{n=0}^{N-1} (u_{n+1} - u_n - \Delta t \, F_n)^2 + W_i(u_0 - I)^2 + w(u_d - U)^2, \tag{4.1.13}$$

where

$$F_n \equiv F(t_n) \quad \text{and} \quad u_d \equiv u(t_d) = u_{n_d}. \tag{4.1.14}$$

Thus the N-point approximation to $\mathcal{J}[u]$ is

$$J_N = J_N(u_0, u_1, \ldots, u_N)$$
$$= J_N(\mathbf{u}), \tag{4.1.15}$$

where

$$\mathbf{u} = (u_0, \ldots, u_N) \in \mathbb{R}^N. \tag{4.1.16}$$

There are very sophisticated ways (Press *et al.*, 1986) to use the information in ∇J_N in order to minimize J_N, but we only need consider the most elementary for now. Suppose we have made k estimates of the minimizer $\hat{\mathbf{u}}$, and the latest is $\mathbf{u}^k$. If $J_N^k \equiv J_N(\mathbf{u}^k)$ is not satisfactorily small, we may naïvely improve on $\mathbf{u}^k$ as follows:

$$\mathbf{u}^{k+1} = \mathbf{u}^k - \theta \nabla J_N^k \tag{4.1.17}$$

for some small positive number θ. Then of course

$$J_N^{k+1} \equiv J_N(\mathbf{u}^{k+1}) = J_N\left(\mathbf{u}^k - \theta \nabla J_N^k\right) \tag{4.1.18}$$

$$\cong J_N(\mathbf{u}^k) - \theta \left|\nabla J_N^k\right|^2 \tag{4.1.19}$$

$$< J_N(\mathbf{u}^k). \tag{4.1.20}$$

Now

$$\left(\nabla J_N^k\right)_n \equiv \frac{\partial J_N}{\partial u_n}(\mathbf{u}^k)$$

$$\cong \frac{J_N\left(u_0^k, u_1^k, \ldots, u_n^k + \Delta u_n, \ldots, u_N^k\right) - J_N(\mathbf{u}^k)}{\Delta u_n}, \tag{4.1.21}$$

where Δu_n is a small increment in u_n, and so we may evaluate (4.1.21) by direct substitution into (4.1.13).

4.1.3   The gradient from the adjoint operator

Evaluation of ∇J_N via (4.1.21) requires N evaluations of J_N, for each of N increments Δu_n, $1 \le n \le N$. As an alternative, we may evaluate the gradient more efficiently by first doing some elementary calculus. It follows easily from (4.1.13) that

$$\frac{\partial J_N}{\partial u_0} = -2W_f\left(\frac{u_1 - u_0}{\Delta t} - F_0\right) + 2W_i(u_0 - I), \tag{4.1.22}$$

$$\frac{\partial J_N}{\partial u_n} = 2W_f\left(\frac{u_n - u_{n-1}}{\Delta t} - F_{n-1} - \frac{u_{n+1} - u_n}{\Delta t} + F_n\right)$$
$$+ 2w\delta_{nn_d}(u_d - U) \tag{4.1.23}$$

for $0 < n < N$, and

$$\frac{\partial J_N}{\partial u_N} = 2W_f\left(\frac{u_N - u_{N-1}}{\Delta t} - F_{N-1}\right). \tag{4.1.24}$$

Hence ∇J_N may be evaluated at roughly the same cost as one evaluation of J_N. Now define the weighted dynamical residual:

$$\lambda_n \equiv W_f \left(\frac{u_{n+1} - u_n}{\Delta t} - F_n \right) \tag{4.1.25}$$

for $0 \leq n \leq N - 1$. Then (4.1.22)–(4.1.24) become

$$\frac{1}{2} \frac{\partial J_N}{\partial u_0} = -\lambda_0 + W_i(u_0 - I), \tag{4.1.26}$$

$$(2\Delta t)^{-1} \frac{\partial J_N}{\partial u_n} = -\frac{\lambda_n - \lambda_{n-1}}{\Delta t} + w \frac{\delta_{nn_d}}{\Delta t}(u_d - U) \tag{4.1.27}$$

for $0 < n < N - 1$, and

$$\frac{1}{2} \frac{\partial J_N}{\partial u_N} = \lambda_{N-1}. \tag{4.1.28}$$

Exercise 4.1.1

Derive the Euler–Lagrange equations for the penalty functional (4.1.5). Discretize as in (4.1.6)–(4.1.12). Compare with (4.1.25)–(4.1.28).   □

The discrete Euler–Lagrange equations are simply

$$\nabla J_N = \mathbf{0}. \tag{4.1.29}$$

In the continuous limit, we might write this as

$$\frac{\delta J[u]}{\delta u(x, t)} = 0. \tag{4.1.30}$$

So our procedure is now:

(i)   estimate $\mathbf{u}^k$;
(ii) evaluate λ^k by substitution of $\mathbf{u}^k$ into (4.1.25);
(iii) evaluate ∇J_N^k by substitution of λ^k into (4.1.26)–(4.1.28); and then
(iv) estimate $\mathbf{u}^{k+1}$ using the gradient information, as in (4.1.17) for example.

4.1.4   "The" variational adjoint method for strong dynamics

There is a special case that has been very widely used ("THE variational adjoint method", e.g., Lewis and Derber, 1985; LeDimet and Talagrand, 1986; Thacker and Long, 1988; Greiner and Perigaud, 1994a,b; Kleeman et al., 1995). Assume that the dynamics are perfect; that is, let

$$W_f / W_i, \ W_f / w \to \infty. \tag{4.1.31}$$

Set

$$\frac{\partial J_N^k}{\partial u_n} = 0 \quad \text{for} \quad 0 < n < N - 1, \tag{4.1.32}$$

and set

$$\frac{\partial J_N^k}{\partial u_N} = 0. \tag{4.1.33}$$

Then λ_n^k may be determined for $0 \leq n \leq N - 1$, by solving (4.1.32) and (4.1.33), that is, by stepping

$$\frac{\lambda_{n-1}^k - \lambda_n^k}{\Delta t} = -w \frac{\delta_{nn_d}}{\Delta t} \left(u_d^k - U \right) \tag{4.1.34}$$

backwards from

$$\lambda_{N-1}^k = 0. \tag{4.1.35}$$

This leads to a value for λ_0^k, and hence for $\frac{\partial J_N^k}{\partial u_0}$ via (4.1.26). Then u_0^{k+1} may be estimated using this remaining amount of nonvanishing gradient information:

$$u_0^{k+1} = u_0^k - \theta \frac{\partial J_N^k}{\partial u_0}. \tag{4.1.36}$$

The rest of $\mathbf{u}^{k+1}$ is evaluated using (4.1.25):

$$\frac{u_{n+1}^{k+1} - u_n^{k+1}}{\Delta t} = F_n + W_f^{-1} \lambda_n^k, \tag{4.1.37}$$

for $0 \leq n < N - 1$. But λ_n is $O(W_i, w)$, and so $W_f^{-1} \lambda_n \to 0$ in the limit as $W_f / W_i \to \infty$, $W_f / w \to \infty$. In other words, each estimate of $\mathbf{u}^k$ obeys the exact or "strong" dynamical constraint

$$\frac{u_{n+1}^k - u_n^k}{\Delta t} = F_n \tag{4.1.38}$$

for $0 \leq n \leq N - 1$. Notice that only the initial value u_0^k is varied independently. The values of $u_1^k, \ldots, u_N^k$ also vary, but they are determined by the value of u_0^k and by (4.1.38). That is, a forward integration is required. A backward integration is also required in order to determine $\boldsymbol{\lambda}^k$.

In practice we deal with partial differential equations, and the initial value u_0^k is a field: $u_0^k = u_0^k(\mathbf{x})$. If the spatial grid is fine, then the initial field u_0^k becomes a long vector, while the time-dependent solution or "state" $(u_1^k(\mathbf{x}), \ldots, u_N^k(\mathbf{x}))$ becomes such a long vector that convergence of descent algorithms is typically very slow. Thus the "weak-constraint" minimization described in §4.1.2 is quite unfeasible, while the "strong-constraint" minimization would seem feasible. Note the imposition of a major physical constraint (exact or "strong" dynamics), in order to deal with a computational difficulty. We may well be able to fit the data fairly closely with an exact solution of a model, but this is of dubious value when, as usual, we have more confidence in the data than in the model. We should instead make estimates of the dynamical error variances and hence the dynamical weight W_f, and then use a minimization technique that is adequate to the task. Furthermore, a reasonable hypothesis is then being tested.

4.1.5   Preconditioning: the Hessian

Convergence towards the minimum of $\mathcal{J}_N$ may be accelerated by the use of second-order information. The Taylor expansion of $J_N(\mathbf{u})$ about the minimum $\hat{\mathbf{u}}$ is

$$J_N(\mathbf{u}) \cong J_N(\hat{\mathbf{u}}) + (\mathbf{u} - \hat{\mathbf{u}})^{\mathrm{T}} \nabla J_N(\hat{\mathbf{u}}) + \frac{1}{2}(\mathbf{u} - \hat{\mathbf{u}})^{\mathrm{T}} \hat{\mathbf{H}}(\mathbf{u} - \hat{\mathbf{u}}) + \cdots, \qquad (4.1.39)$$

where the components of the Hessian matrix $\hat{\mathbf{H}}$ are

$$\hat{H}_{nm} = \frac{\partial^2 J_N}{\partial u_n \partial u_m}(\hat{\mathbf{u}}), \qquad (4.1.40)$$

for $1 \leq n, m \leq N$. Since $\hat{\mathbf{u}}$ is an extremum of J_N,

$$\nabla J_N(\hat{\mathbf{u}}) = \mathbf{0}, \qquad (4.1.41)$$

hence

$$J_N(\mathbf{u}) \cong J_N(\hat{\mathbf{u}}) + \frac{1}{2}(\mathbf{u} - \hat{\mathbf{u}})^{\mathrm{T}} \hat{\mathbf{H}}(\mathbf{u} - \hat{\mathbf{u}}). \qquad (4.1.42)$$

So, near $\hat{\mathbf{u}}$, the gradient of J_N is approximately given by

$$\nabla J_N(\mathbf{u}) \cong \hat{\mathbf{H}}(\mathbf{u} - \hat{\mathbf{u}}). \qquad (4.1.43)$$

Assuming that $\hat{\mathbf{H}}$ is invertible,

$$\hat{\mathbf{H}}^{-1} \nabla J_N(\mathbf{u}) \cong \mathbf{u} - \hat{\mathbf{u}}. \qquad (4.1.44)$$

If we step in the direction of (4.1.44), then

$$\mathbf{u}^{k+1} = \mathbf{u}^k - \theta (\mathbf{H}^k)^{-1} \nabla J_N^k \cong \mathbf{u}^k - \theta(\mathbf{u}^k - \hat{\mathbf{u}}). \qquad (4.1.45)$$

The advantage of this approach is clear once we examine Fig. 2.1.1. If the Hessian is poorly conditioned, that is, if its eigenvalues have a large range, then contours of constant J_N will be highly eccentric ellipsoids. A step from $\mathbf{u}^k$ down the gradient ∇J_N^k will not be a step towards the minimum at $\hat{\mathbf{u}}$. However, a step from $\mathbf{u}^k$ in the direction $(\hat{\mathbf{H}}^k)^{-1} \nabla J_N^k$ will be a step approximately towards $\hat{\mathbf{u}}$, and so convergence towards $\hat{\mathbf{u}}$ should be more rapid.

It would seem that such "preconditioning" would be unfeasible if N is very large, since it would be so expensive to compute and invert the $N \times N$ matrix $\mathbf{H}$ at each step. However (Le Dimet et al., 1997), $\mathbf{H}^{-1} \nabla J$ may be evaluated iteratively without first computing $\mathbf{H}$, just as $\mathbf{P}^{-1}\mathbf{h}$ may be evaluated iteratively without first computing $\mathbf{P}$ (see §3.1.4). A final caution is that $\mathbf{H}$ is $N \times N$, where N is the number of state variables, while $\mathbf{P}$ is $M \times M$, where M is the number of data.

Note. Inspecting (2.1.14) and (2.1.17) shows that

$$\mathbf{S} \equiv \mathbf{R} + \mathbf{R}\,\mathbf{C}_\epsilon^{-1}\mathbf{R} = \mathbf{P}\,\mathbf{C}_\epsilon^{-1}\,\mathbf{P} - \mathbf{P}, \qquad (4.1.46)$$

where

$$S_{ij} = \frac{\partial^2 \mathcal{J}}{\partial \beta_i \partial \beta_j}. \qquad (4.1.47)$$

Thus $\mathbf{R}$ and $\mathbf{P}$ are closely related to $\mathbf{S}$, the Hessian of the penalty functional $\mathcal{J}[u]$ with respect to the observable degrees of freedom.

For a review of the use of "adjoint models" see Errico (1997).

4.1.6   Continuous adjoints or discrete adjoints?

The derivation of gradients and Euler–Lagrange equations for penalty functions was illustrated in §4.1.2 with a trivial example. (Note that J_N defined by (4.1.13) is a real-valued *function* of the N-dimensional vector $\mathbf{u} = (u_1, \ldots, u_N)^{\mathrm{T}}$, whereas $\mathcal{J}$ defined by (4.1.5) is a real-valued *functional* of the function $u = u(t)$.) Derivation of "discrete adjoints" becomes exacting and tedious as the complexity of the forward numerical model increases. Is it worth the trouble? After all, one could with comparative ease derive the adjoint operators analytically and then approximate numerically as in the forward model. In general, proceeding in that order "breaks" adjoint symmetry; the "discrete adjoint equation" is not the "adjoint discrete equation". The broken symmetry manifests itself directly as a spurious, asymmetric part in the representer matrix. So long as the asymmetric part is relatively small, it could be discarded. The resulting inverse solution would be slightly suboptimal. In the indirect representer algorithm of §3.1.4, the representer matrix is not being explicitly constructed, hence its asymmetry cannot be suppressed. A preconditioned biconjugate-gradient solver must be used in the iterative search for the representer coefficients. It seems cleaner to work with the "adjoint discrete equation"; then asymmetry of the representer matrix becomes a very useful indicator of coding errors.

Again, deriving the adjoint discrete equation is an exacting task. Experience and technique are important. Recognition of pattern can greatly reduce the burden. As a rule, centered finite-difference operators are self-adjoint. Most difficulties occur at boundaries, where operators are typically one-sided and so not self-adjoint. The introduction of virtual state variables outside the domain of the forward model can simplify the resulting adjoint discrete equation (J. Muccino, personal communication). For example, consider the simple conduction problem

$$\frac{\partial T}{\partial t} = \theta \frac{\partial^2 T}{\partial x^2} \qquad (4.1.48)$$

for constant $\theta > 0$, for $0 \le x \le x_{\max}$ and $0 \le t \le t_{\max}$, subject to the initial condition

$$T(x, 0) = I(x) \qquad (4.1.49)$$

for $0 \leq x \leq x_{\max}$, and the heat-reservoir boundary conditions

$$T(0, t) = L(t), \quad T(x_{\max}, t) = R(t) \tag{4.1.50}$$

for $0 \leq t \leq t_{\max}$. A simple, explicit time-stepping scheme is provided by

$$T_m^{k+1} - T_m^k = (\theta \Delta t / \Delta x^2)(T_{m+1}^k + T_{m-1}^k - 2T_m^k), \tag{4.1.51}$$

where

$$T_m^k = T(m\Delta x, k\Delta t), \quad 0 \leq m \leq M, \quad 0 \leq k \leq K, \quad \Delta x = x_{\max}/M, \quad \Delta t = t_{\max}/K. \tag{4.1.52}$$

It is straightforward to devise a penalty function for generalized inversion of the well-posed discrete forward problem corresponding to (4.1.48)–(4.1.50) plus an over-determining set of temperature data. Inspection of (4.1.51) indicates that the spatial summation of weighted squares of residuals in (4.1.51) should range from $m = 1$ to $m = M - 1$. The resulting Euler–Lagrange equations are inelegant, but improve with the introduction of bogus weighted residuals λ_0^k and λ_M^k that are identically zero. Alternatively, virtual temperatures T_{-1}^k and T_{M+1}^k may be introduced. Residuals λ_0^k and λ_M^k are now automatically defined, and the spatial summations of squared residuals should be extended to $0 \leq m \leq M$. The Euler–Lagrange equations are then far tidier. In particular, variation with respect to T_{-1}^k and T_{M+1}^k implies that λ_0^k and λ_M^k vanish.

Practitioners of spectral methods or finite element methods are free from all these vexed considerations. Derivatives are transferred to the basis functions or to the elements by partial integration, and are then evaluated analytically.

4.2   The sweep algorithm, sequential estimation and the Kalman filter

4.2.1   More trickery from control theory

The Euler–Lagrange equations are a two-point boundary-value problem in the time interval of interest. We used representers in order to untangle this problem, that is, in order to express it in terms of initial value problems. The Gelfand and Fomin sweep algorithm provides a remarkable alternative (Gelfand and Fomin, 1963; Meditch, 1970). Partial implementation of the algorithm yields an appealing "sequential estimation" scheme for assimilating data. This control-theoretic derivation of the Kalman filter follows logically from preceding chapters, but the reader may prefer to start with the statistical derivation of the Kalman filter in §4.3. That is, §4.2 may be omitted from a first reading.

4.2.2   The sweep algorithm yields the Kalman filter

Let us gather up all the parts of the Euler–Lagrange system for our simple model (1.2.6)–(1.2.8) and for the penalty functional (1.5.7), in the most general form that we have used:

$$-\frac{\partial \lambda}{\partial t}(x,t) - c\frac{\partial \lambda}{\partial x}(x,t) = (\mathbf{d}-\hat{\mathbf{u}})^{\mathrm{T}}\mathbf{C}_\epsilon^{-1}\boldsymbol{\delta}, \tag{4.2.1) = (1.3.1}$$

$$\lambda(x,T) = 0, \tag{4.2.2) = (1.3.2}$$

$$\lambda(L,t) = 0, \tag{4.2.3) = (1.3.3}$$

$$\frac{\partial \hat{u}}{\partial t}(x,t) + c\frac{\partial \hat{u}}{\partial x}(x,t) = F(x,t) + (C_f \bullet \lambda)(x,t), \tag{4.2.4) = (1.5.14}$$

$$\hat{u}(x,0) = I(x) + (C_i \circ \lambda)(x), \tag{4.2.5) = (1.5.15}$$

$$\hat{u}(0,t) = B(t) + c(C_b * \lambda)(t). \tag{4.2.6) = (1.5.16}$$

Recall that the covariances C_f, C_i, C_b and $\mathbf{C}_\epsilon$ are the inverses of the weights W_f, W_i, W_b and $\mathbf{w}$: see §1.5. The impulse vector $\boldsymbol{\delta}$ has, as its mth component, the scalar impulse $\delta(x-x_m)\delta(t-t_m)$.

First, assume that the dynamical residual f is uncorrelated in time:

$$C_f(x,y,t,s) = Q_f(x,y)\delta(t-s), \tag{4.2.7}$$

and so the "forward" Euler–Lagrange equation (4.2.4) becomes

$$\frac{\partial \hat{u}}{\partial t}(x,t) + c\frac{\partial \hat{u}}{\partial x}(x,t) = F(x,t) + \int_0^L dy\, Q_f(x,y)\lambda(y,t). \tag{4.2.8}$$

Next, assume that the inverse estimate $\hat{u}$ is linearly related to the weighted residual or adjoint variable λ:

$$\hat{u}(x,t) = \int_0^L dy\, P(x,y,t)\lambda(y,t) + v(x,t), \tag{4.2.9}$$

for some "slope" P and "intercept" v. With the substitution of (4.2.9), the left-hand side of (4.2.8) becomes

$$\int_0^L dy \left\{ \frac{\partial P}{\partial t}(x,y,t)\lambda(y,t) + P(x,y,t)\frac{\partial \lambda}{\partial t}(y,t) + c\frac{\partial P}{\partial x}(x,y,t)\lambda(y,t) \right\}$$

$$+ \frac{\partial v}{\partial t}(x,t) + c\frac{\partial v}{\partial x}(x,t). \tag{4.2.10}$$

Now assume that the measurement errors ϵ are uncorrelated, if at different times:

$$\overline{\epsilon_n \epsilon_m} = 0 \quad \text{if} \quad t_n \neq t_m \tag{4.2.11}$$

for $1 \leq n, m \leq M$, in which case the $M \times M$ measurement error covariance matrix consists of K blocks on the diagonal, where each block is an $N \times N$ matrix; K is the number of measurement times, N is the number of measurement sites, and $M = NK$. That is, $\mathbf{C}_\epsilon = \mathrm{diag}\,(\mathbf{C}_\epsilon^1, \ldots, \mathbf{C}_\epsilon^K)$. Then "the" Euler–Lagrange equation (4.2.1) becomes

$$-\frac{\partial \lambda}{\partial t}(x, t) - c\frac{\partial \lambda}{\partial x}(x, t) = \sum_{k=1}^{K}(\mathbf{d}^k - \hat{\mathbf{u}}^k)^{\mathrm{T}}(\mathbf{C}_\epsilon^k)^{-1}\delta^k, \tag{4.2.12}$$

where $(\mathbf{d}^k)_n$ is the datum at site x_n at time t_k, while

$$(\delta^k)_n = \delta(x - x_n)\delta(t - t_k) \tag{4.2.13}$$

and, according to the substitution (4.2.9),

$$(\hat{\mathbf{u}}^k)_n \equiv \hat{u}(x_n, t_k) = \int_0^L dy\, P(x_n, y, t_k)\lambda(y, t_k) + v(x_n, t_k). \tag{4.2.14}$$

Exercise 4.2.1
Now substitute (4.2.14) into the rhs of (4.2.12), and then (4.2.12) into (4.2.10), which is the lhs of (4.2.8). At times $t \neq t_k$, $1 \leq k \leq K$, integrate over y by parts and equate coefficients of $\lambda(y, t)$, yielding

$$\frac{\partial P}{\partial t}(x, y, t) + c\frac{\partial P}{\partial x}(x, y, t) + c\frac{\partial P}{\partial y}(x, y, t)$$

$$+ cP(x, 0, t)\delta(y) = Q_f(x, y), \tag{4.2.15}$$

and leaving

$$\frac{\partial v}{\partial t}(x, t) + c\frac{\partial v}{\partial x}(x, t) = F(x, t). \tag{4.2.16}$$

$$\square$$

From the initial condition (4.2.5), we may analogously obtain

$$P(x, y, 0) = C_i(x, y) \tag{4.2.17}$$

and

$$v(x, 0) = I(x). \tag{4.2.18}$$

If we assume that the errors in the boundary data are uncorrelated in time:

$$C_b(t, s) \equiv \overline{b(t)b(s)} = Q_b\delta(t - s), \tag{4.2.19}$$

then (4.2.6) becomes

$$\hat{u}(0, t) = B(t) + cQ_b\lambda(0, t). \tag{4.2.20}$$

We recover

$$P(0, y, t) = cQ_b\delta(y) \tag{4.2.21}$$

and

$$v(0, t) = B(t). \tag{4.2.22}$$

The system (4.2.15), (4.2.17), (4.2.21) and the system (4.2.16), (4.2.18), (4.2.22) may be integrated forward in time until $t = t_1-$. Note that the initial value and forcing for $P(x, y, t)$, respectively $C_i(x, y)$ and $Q_f(x, y)$, are symmetric. If we assume that $P(x, y, t)$ is symmetric, then by (4.2.21)

$$P(x, 0, t) = P(0, x, t) = cQ_b\delta(x), \tag{4.2.23}$$

and hence the seemingly symmetry-breaking fourth term on the lhs of (4.2.15) is

$$cP(x, 0, t)\delta(y) = c^2 Q_b\delta(x)\delta(y), \tag{4.2.24}$$

which is symmetric. In other words, assuming symmetry leads to no contradiction.

It remains to determine the jumps in P and v as t passes through t_k. First, we learn from (4.2.12) that λ is discontinuous in t_k, with

$$-\lambda(x, t_k+) + \lambda(x, t_k-) = (\mathbf{d}^k - \hat{\mathbf{u}}^k)^{\mathrm{T}}(\mathbf{C}_\epsilon^k)^{-1}\boldsymbol{\delta}, \tag{4.2.25}$$

where $(\boldsymbol{\delta})_n = \delta(x - x_n)$. We infer from (4.2.4) that $\hat{u}$ is continuous at t_k. Hence (4.2.9) implies that

$$\left[\int_0^L dy\, P(x, y, t)\lambda(y, t) + v(x, t)\right]_{t_k-}^{t_k+} = 0. \tag{4.2.26}$$

Substitute (4.2.14) into (4.2.25) and then substitute (4.2.25) into (4.2.26). Equating terms proportional to $\lambda(x, y, t_k-)$ yields, after a little algebra,

$$\begin{aligned} P(x, y, t_k+) &- P(x, y, t_k-) \\ &= -\mathbf{P}^{k-}(x)^{\mathrm{T}}(\mathbf{P}^{k-} + \mathbf{C}_\epsilon^k)^{-1}\mathbf{P}^{k-}(y) \end{aligned} \tag{4.2.27}$$

and

$$v(x, t_k+) - v(x, t_k-) = \mathbf{K}^{k^{\mathrm{T}}}(\mathbf{d}^k - \mathbf{v}^{k-}), \tag{4.2.28}$$

where

$$P_n^{k-}(x) \equiv P(x_n, x, t_k-) \tag{4.2.29}$$

$$P_{nm}^{k-} \equiv P(x_n, x_m, t_k-) \tag{4.2.30}$$

$$\mathbf{K}^k \equiv (\mathbf{C}_\epsilon^k)^{-1}\mathbf{P}^{k+} \tag{4.2.31}$$

and

$$(\mathbf{v}^{k-})_n = v(x_n, t_k-). \tag{4.2.32}$$

We now have an explicit algorithm for P and v, for all $t \geq 0$. To complete the formula (4.2.9) for $\hat{u}$, we need λ. Now λ obeys (4.2.1)–(4.2.3), but (4.2.1) involves $\hat{u}$. We may eliminate $\hat{u}$ using (4.2.14), yielding an equation for P, v and λ. We can determine P and v, so λ may be found by backwards integration, and the Gelfand and Fomin sweep is complete.

Exercise 4.2.2

Derive the equation for λ, free of $\hat{u}$. ☐

Note 1. The above procedure has a major drawback: it would be necessary to compute and store $P(x, y, t)$ and $v(x, t)$ for $0 < x, y < L$, and for $0 < t < T$. This would be prohibitive in practice.

Note 2. It is only necessary to store $P(x, y, t_k+)$ and $v(x, t_k+)$ in order to evaluate $\hat{u}^k$, for $1 \le k \le K$, and hence λ (see(4.2.12)). Having solved for λ, we could then find $\hat{u}$ by integrating (4.2.4)–(4.2.6).

Note 3. There are other such "control theory" algorithms such as that of Rauch, Tung and Streibel (e.g., Gelb, 1974), but these require even more computation and storage. These algorithms are impractical for the generalized inversion of oceanic or atmospheric models.

Note 4. The adjoint variable λ vanishes after assimilating the last data: $\lambda = 0$ for $t_K < t < T$, hence the generalized inverse $\hat{u}$ agrees with the "intercept" v at the end of the smoothing interval:

$$\hat{u}(x, t) = v(x, t) \tag{4.2.33}$$

for $t_K < t < T$, where T is somewhat arbitrary. So, if we only want to know the influence of the Kth (the latest) data $\mathbf{d}^K$ upon the circulation estimate $\hat{u}$ at time t_K (the present), then we need not do more than solve for v (which requires solving for P: see (4.2.28)–(4.2.32)). The previous data: $\mathbf{d}^1, \ldots, \mathbf{d}^{K-1}$ also influence v at time t_K, but $\mathbf{d}^K$ has no influence on v for $t < t_K$. Thus v is a "sequential" estimate of u, using data as they arrive.

4.3   The Kalman filter: statistical theory

4.3.1   Linear regression

The Kalman filter has just been derived as a first step in solving linear Euler–Lagrange problems. It is a sequential algorithm, that is, it calculates the generalized inverse at times later than all the data: $v(x, t) = \hat{u}(x, t)$ for all $t > t_K$. Recall that $\hat{u}$ minimizes a quadratic penalty functional over $0 \le t \le T$, where $t_K < T$. The Kalman filter will now be derived using linear regression.

4.3.2   Random errors: first and second moments

Our ocean model is

$$\frac{\partial u}{\partial t} + c \frac{\partial u}{\partial x} = F + f, \tag{4.3.1}$$

for $0 \leq x \leq L$ and $0 \leq t \leq T$, subject to the boundary condition

$$u(0, t) = B(t) + b(t) \qquad (4.3.2)$$

and the initial condition

$$u(x, 0) = I(x) + i(x). \qquad (4.3.3)$$

We have assumed that F, B and I are unbiased estimates of the forcing, boundary and initial values:

$$Ef = Eb = Ei = 0, \qquad (4.3.4)$$

and we prescribed the autocovariances of f, b and i:

$$E(f(x, t)f(y, s)) = Q_f(x, y)\delta(t - s), \qquad (4.3.5)$$

$$E(b(t)b(s)) = Q_b\delta(t - s), \qquad (4.3.6)$$

$$E(i(x)i(y)) = C_i(x, y). \qquad (4.3.7)$$

We assumed that their cross-covariances all vanish: $E(fb) = E(fi) = E(bi) = 0$.

There are data at N points $x_1, \ldots, x_N$, at discrete times $t = t_1, \ldots, t_K$:

$$d_n^k = u(x_n, t_k) + \epsilon_n^k \qquad (4.3.8)$$

for $1 \leq n \leq N$, where ϵ_n^k are the measurement errors, for which

$$E\epsilon^k = E(f\epsilon^k) = E(b\epsilon^k) = E(i\epsilon^k) = \mathbf{0}, \qquad (4.3.9)$$

$$E(\epsilon^k \epsilon^{lT}) = \delta_{kl}\mathbf{C}_\epsilon^k. \qquad (4.3.10)$$

That is, f, b and ϵ^k are uncorrelated in time. The vectors in (4.3.9), (4.3.10) have N components. Note that the points $x_1, \ldots, x_N$ do not necessarily coincide with a spatial grid for numerical integration of (4.3.1)–(4.3.3); they are merely a set of N measurement sites.

4.3.3   Best linear unbiased estimate: before data arrive

We shall now construct $w(x, t)$, the best linear unbiased estimate of $u(x, t)$, given data prior to t. Assuming $t_1 > 0$, at time $t = 0$ we can do no better than

$$w(x, 0) = I(x), \qquad (4.3.11)$$

for which the error variance is C_i: see (4.3.7). For $0 \leq t \leq t_1-$, let

$$\frac{\partial w}{\partial t} + c\frac{\partial w}{\partial x} = F, \qquad (4.3.12)$$

$$w(0, t) = B(t). \qquad (4.3.13)$$

The error $e \equiv u - w$ obeys

$$\frac{\partial e}{\partial t}(x, t) + c\frac{\partial e}{\partial x}(x, t) = f(x, t), \qquad (4.3.14)$$

$$e(x, 0) = i(x), \qquad (4.3.15)$$

$$e(0, t) = b(t), \qquad (4.3.16)$$

the solution of which is

$$e(x, t) = \int_0^t \int_0^L ds\, d\xi\, \gamma(\xi, s, x, t) f(\xi, s) + c\int_0^t ds\, \gamma(0, s, x, t)b(s)$$

$$+ \int_0^L d\xi\, \gamma(\xi, 0, x, t)i(\xi), \qquad (4.3.17)$$

where γ is the Green's function (see §1.1.4). Hence $E(e(x, t)f(y, t)) = \frac{1}{2}Q_f(x, y)$, since $\gamma(x, t, y, t) = \delta(x - y)$, and $\int_{t-}^t \delta(s)\, ds = \frac{1}{2}$. Also

$$E(e(x, t)b(t)) = cQ_b\delta(x). \qquad (4.3.18)$$

Now define the spatial *error* covariance at time t by

$$P(x, y, t) \equiv E(e(x, t)e(y, t)) = P(y, x, t). \qquad (4.3.19)$$

Multiplying (4.3.14) by $e(y, t)$ and averaging yields

$$\frac{\partial P}{\partial t}(x, y, t) + c\frac{\partial P}{\partial x}(x, y, t) + c\frac{\partial P}{\partial y}(x, y, t) = Q_f(x, y); \qquad (4.3.20)$$

multiplying (4.3.16) by $e(y, t)$ and averaging yields

$$P(0, y, t) = cQ_b\delta(y), \qquad (4.3.21)$$

$$P(x, 0, t) = cQ_b\delta(x). \qquad (4.3.22)$$

Initially,

$$P(x, y, 0) = C_i(x, y). \qquad (4.3.23)$$

4.3.4   Best linear unbiased estimate: after data have arrived

The situation at time t_1- is that we have an estimate $w_-^1(x) \equiv w(x, t_1-)$, equal to the mean of $u(x, t_1)$, and we have its *error* covariance $P_-^1(x, y) \equiv P(x, y, t_1-)$. The new information are the data $\mathbf{d}^1$. These too contain random errors, but by (4.3.8) we are assuming that

$$E\mathbf{d}^1 = E\mathbf{u}^1. \qquad (4.3.24)$$

Let us seek a new estimate $w_+^1(x)$ for $u(x, t_1)$ which is linear in $w_-^1(x)$ and associated data misfits:

$$w_+^1(x) = \alpha w_-^1(x) + \mathbf{s}(x)^{\mathrm{T}}(\mathbf{d}^1 - \mathbf{w}_-^1), \qquad (4.3.25)$$

where

$$(\mathbf{w}_-^1)_n = w_-^1(x_n) \equiv w(x_n, t_1-).$$ (4.3.26)

The constant α and the interpolant $\mathbf{s}(x)$ have yet to be chosen. Consider the error

$$e_+^1(x) = u(x, t_1) - w_+^1(x).$$ (4.3.27)

Now

$$u(x, t_1) = w_-^1(x) + e_-^1(x),$$ (4.3.28)

where $e_-^1(x) = e(x, t_1-)$. Hence

$$e_+^1(x) = (1 - \alpha)w_-^1(x) + e_-^1(x) - \mathbf{s}(x)^{\mathrm{T}}(\mathbf{d}^1 - \mathbf{w}_-^1).$$ (4.3.29)

But $Ee_-^1 = 0$. So if we choose $\alpha = 1$, then

$$Ee_+^1(x) = 0$$ (4.3.30)

and (4.3.25) is an unbiased estimate. The error variance is

$$P_+^1(x, x) = E\left(e_+^1(x)^2\right).$$ (4.3.31)

Exercise 4.3.1

Show that the error variance (4.3.31) is least if the optimal interpolant is

$$\mathbf{s}(x) = \mathbf{K}^1(x),$$ (4.3.32)

where the "Kalman gain" vector field $\mathbf{K}^1(x)$ in (4.3.32) is

$$\mathbf{K}^1(x) = \left[\mathbf{P}_-^1 + \mathbf{C}_\epsilon^1\right]^{-1}\mathbf{P}_-^1(x).$$ (4.3.33)

The vector $\mathbf{P}_-^1(x)$ and matrix $\mathbf{P}_-^1$ have components

$$P_{n-}^1(x) = P_-^1(x_n, x),$$

$$P_{nm-}^1 = P_-^1(x_n, x_m).$$ (4.3.34)

□

Exercise 4.3.2

Show that the posterior error covariance at time t_1 is

$$P_+^1(x, y) \equiv E\left(e_+^1(x)e_+^1(y)\right)$$
$$= P_-^1(x, y) - \mathbf{P}_-^1(x)^{\mathrm{T}}\left[\mathbf{P}_-^1 + \mathbf{C}_\epsilon^1\right]^{-1}\mathbf{P}_-^1(y).$$ (4.3.35)

□

Clearly, we may repeat this construction at $t_2, t_3, \ldots$. See Fig. 4.3.1.
Gathering up all the results, the Kalman filter estimate w satisfies

$$\frac{\partial w}{\partial t} + c\frac{\partial w}{\partial x} = F$$ (4.3.36)

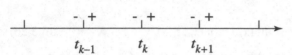

Figure 4.3.1 Time line for the Kalman filter.

for $0 \leq x \leq L$, $t_k < t < t_{k+1}$, subject to

$$w(x, 0) = I(x) \tag{4.3.37}$$

and

$$w(0, t) = B(t). \tag{4.3.38}$$

The change in w at time t_k is

$$w(x, t_k+) - w(x, t_k-) = \mathbf{K}^k(x)^{\mathrm{T}}(\mathbf{d}^k - \mathbf{w}_-^k), \tag{4.3.39}$$

where the Kalman gain is

$$\mathbf{K}^k(x) = \left[\mathbf{P}_-^k + \mathbf{C}_\epsilon^k\right]^{-1}\mathbf{P}_-^k(\mathbf{x}). \tag{4.3.40}$$

The error covariance satisfies

$$\frac{\partial P}{\partial t} + c\frac{\partial P}{\partial x} + c\frac{\partial P}{\partial y} = Q_f \tag{4.3.41}$$

for $0 \leq x, y \leq L$, $t_k < t < t_{k+1}$, subject to

$$P(0, y, t) = cQ_b\delta(y), \tag{4.3.42}$$

$$P(x, 0, t) = cQ_b\delta(x) \tag{4.3.43}$$

and

$$P(x, y, 0) = C_i(x, y). \tag{4.3.44}$$

The change in P at t_k is

$$P(x, y, t_k+) - P(x, y, t_k-) = -\mathbf{P}_-^k(x)^{\mathrm{T}}\mathbf{K}^k(y). \tag{4.3.45}$$

The new data always reduce the error variance at data sites. Note carefully the assumptions that the dynamical and boundary errors f, b and the data ϵ^k are uncorrelated in time, and that the different types of errors are not cross-correlated. Note also that the optimal choices for α and $s(x)$ in (4.3.25) are not random. They depend not upon the random inputs $w_-^1(x)$, $\mathbf{d}^1$ but upon the covariances of the errors in the inputs.

4.3.5   Strange asymptotics

It is usually assumed that the data errors are statistically stationary, that is, $\mathbf{C}_\epsilon^k$ is independent of k. It is often the case that the temporal sampling interval $t_{k+1} - t_k$ also is independent of k. Consequently, the Kalman filter error covariance P approaches an equilibrium state, in which $P(x, y, t_k-) = P(x, y, t_{k+1}-)$ and $P(x, y, t_k+) = P(x, y, t_{k+1}+)$. The covariance does still evolve in time from t_k+ to $t_{k+1}-$, but Q_f and Q_b are independent of t and so the evolution is the same in every data interval. In general we are interested in more complicated dynamics than are expressed in (4.3.1); so long as the dynamics are linear, they may be expressed as

$$\frac{\partial u}{\partial t} + L_x u = F + f, \tag{4.3.46}$$

where L_x is a linear partial differential operator with respect to x. Of course, Primitive Equation models involve many dependent variables, but we shall retain just one here, namely u, for clarity. The error covariance now satisfies

$$\frac{\partial P}{\partial t}(x, y, t) + L_x P(x, y, t) + L_y P(x, y, t) = Q_f(x, y). \tag{4.3.47}$$

To simplify the discussion further, let us assume that the data interval $\Delta t = t_{k+1} - t_k$ is much smaller than the evolution time scale for (4.3.47), so that (see Fig. 4.3.1)

$$P_- = P_+ - \Delta t(L_x P_- + L_y P_-) + \Delta t Q_f + O(\Delta t^2), \tag{4.3.48}$$

where $P_- = P(x, y, t_{k+1}-)$ and $P_+ = P(x, y, t_k+)$. Recall that both P_+ and P_- are independent of k at equilibrium, hence (4.3.40) and (4.3.45) yield

$$P_+ = P_- - P_-^{\mathrm{T}}(P_- + C_\epsilon)^{-1}P_-. \tag{4.3.49}$$

Combining (4.3.48) and (4.3.49) we have

$$\Delta t(L_x P_- + L_y P_-) + P_-^{\mathrm{T}}(P_- + C_\epsilon)^{-1}P_- = \Delta t Q_f + O(\Delta t^2). \tag{4.3.50}$$

Notice the nonlinearity of the impact of data sites upon P. It is possible for P to strike a balance between the two terms on the left-hand side of (4.3.50) (dynamics and data-impact). This balance can take the form of a boundary layer around data sites. The Kalman gain $\mathbf{K}$ and the Kalman filter estimate w will have this structure, which is quite unphysical (Bennett, 1992). It arises from the adoption of a "cycling" algorithm, as in (4.3.45).

Exercise 4.3.3
Show that there is no such nonlinearity in the non-sequential representer algorithm, for one fixed smoothing interval $[0, T]$ that may include many measurement times: $0 < t_1 < \cdots < t_n < \cdots < t_N < T$.   □

Exercise 4.3.4
Consider smoothing a sequence of such intervals: $KT < t < (K + 1)T$, $K = 0, 1, 2, \ldots$, using the inverse estimate at the end of the Kth interval as the first-guess initial field at the start of the $(K + 1)^{\mathrm{th}}$: $I^{K+1}(x) = \hat{u}^K(x, (K + 1)T)$, and using the error covariance for the inverse estimate as the error covariance for the first-guess initial field: $C_i^{K+1}(x, y) = C_{\hat{u}}^K(x, y, (K + 1)T)$. Show that the equilibrium error covariance for this "cycling" inverse obeys a nonlinear equation like (4.3.50). Hint: for simplicity,

assume that the domain is infinite: $-\infty < x < \infty$, assume that the first-guess forcing field F^K is perfect: $C_f^K = 0$, and integrate (3.2.43) as crudely as (4.3.48).   □

4.3.6 "Colored noise": the augmented Kalman filter

We may relax the assumption (4.3.5) of "white system noise". The simplest "colored system noise" has covariance

$$E(f(x,t)f(y,s)) = Q_f(x,y)e^{-\frac{|t-s|}{\tau}} \qquad (4.3.51)$$

for some decorrelation time scale $\tau > 0$. Note that the Q_fs appearing in (4.3.5) and (4.3.51) have different units of measurement. It may be shown that (4.3.51) is satisfied by solutions of the ordinary differential equation

$$\frac{df}{dt}(x,t) - \tau^{-1}f(x,t) = q(x,t), \qquad (4.3.52)$$

provided

$$E(q(x,t)q(y,s)) = (\tau/2)^{-1}Q_f(x,y)\delta(t-s), \qquad (4.3.53)$$

$$E(f(x,0)f(y,0)) = Q_f(x,y), \qquad (4.3.54)$$

and

$$E(f(x,0)q(y,s)) = 0. \qquad (4.3.55)$$

This suggests *augmenting* the state variable (Gelb, 1974):

$$u(x,t) \rightarrow \begin{pmatrix} u(x,t) \\ f(x,t) \end{pmatrix}. \qquad (4.3.56)$$

The dynamical model is now (4.3.46), (4.3.52). Note that the "colored" random process $f(x,t)$ is now part of the state to be estimated. The augmented system is driven by the "white noise" $q(x,t)$. The augmented error covariance now includes cross-covariances of errors in the Kalman filter estimates of u and f.

4.3.7 Economies

The Kalman filter is a very popular data assimilation technique, owing to its being sequential (e.g., Fukumori and Malanotte-Rizzoli, 1995; Fu and Fukumori, 1996; Chan *et al.*, 1996). Also, the "analysis" step (4.3.39) is identical to synoptic or spatial optimal interpolation, as widely practiced already in meteorology and oceanography (Miller, 1996; Malanotte-Rizzoli *et al.*, 1996; Hoang *et al.*, 1997a; Cohn, 1997). The Kalman filter algorithm evolves the error covariance P in time, via (4.3.41), and (4.3.45). Nevertheless, evolving P is a massive task for realistically large systems so many compromises are made. For example, the covariance $P(x,y,t)$ is evolved on a computational grid much coarser than the one used for the state estimate $w(x,t)$, or $P(x,y,t)$

is integrated to an equilibrium covariance $P_\infty(x, y)$ which is then used at all times $t_1, \ldots, t_K$ (Fukumori and Malanotte-Rizzoli, 1995), or the number of degrees of freedom in $w(x, t)$ is reduced by an expansion in spatial modes (Hoang et al., 1997b). A covariance such as P may also be approximated by statistical simulation, as discussed in §3.2.

4.4   Maximum likelihood, Bayesian estimation, importance sampling and simulated annealing

4.4.1   NonGaussian variability

Least-squares is the simplest of all estimators. It has so many merits. Gradients and Euler–Lagrange equations are available, so long as the dynamics are smooth. Structural analyses in terms of null spaces, data spaces, representers and sweep algorithms are available, as are statistical closures such as the Kalman filter, when the dynamics are linear or linearizable. Why, then, choose other estimators? Consider ocean temperatures near the Gulf Stream front. As the latter meanders back and forth across the mooring, the temperature switches rapidly between the higher value for the warm Sargasso Sea water and the lower value for the cool slope water. Thus the frequency distribution of temperature would be bimodal, with peaks at the two values. A least-squares analysis of temperature would yield the average temperature, which is in fact realized only briefly while the front is passing through the mooring. What would be a more suitable estimator? Can samples of the non-normal population be generated? How can its estimator be minimized?

4.4.2   Maximum likelihood

Let us review some introductory statistics. Suppose the continuous random variable u has the probability distribution function $p(u; \theta)$, where θ is some parameter. Let $u_1, \ldots, u_n$ be independent samples of u. Then the joint pdf of the samples is the *likelihood function*:

$$L(\theta) = p(u_1, \ldots, u_n; \theta) = \prod_{i=1}^{n} p(u_i; \theta). \tag{4.4.1}$$

That is, $L \prod_{i=1}^{n} du_i$ is the probability that the n samples are in the respective intervals $(u_i, u_i + du_i)$, $1 \le i \le n$. The maximum likelihood estimate of θ is that value of θ for which $L(\theta)$ assumes its maximum value.

As an illustration, suppose that u is normally distributed with mean μ and variance σ^2:

$$p(u; \mu, \sigma) = (2\pi\sigma^2)^{-\frac{1}{2}} \exp[-(2\sigma^2)^{-1}(u - \mu)^2]. \tag{4.4.2}$$

Note that there are two parameters here: μ and σ. Given n samples of u, what are the maximum likelihood estimates of μ and σ^2? The likelihood function is

$$L(\mu, \sigma) \equiv \prod_{i=1}^{n} p(u_i; \mu, \sigma) \tag{4.4.3}$$

$$= (2\pi\sigma^2)^{-n/2} \exp\left[-(2\sigma^2)^{-1} \sum_{i=1}^{n}(u_i - \mu)^2\right]. \tag{4.4.4}$$

We may as well seek the maximum of

$$l = \log L = -\frac{n}{2}\log(2\pi\sigma^2) - (2\sigma^2)^{-1}\sum_{i=1}^{n}(u_i - \mu)^2. \tag{4.4.5}$$

Extremal conditions are

$$\frac{\partial l}{\partial \mu} = -2(2\sigma^2)^{-1}\sum_{i=1}^{n}(u_i - \mu) = 0, \tag{4.4.6}$$

$$\frac{\partial l}{\partial(\sigma^2)} = -\frac{n}{2}\sigma^{-2} + \frac{1}{2}\sigma^{-4}\sum_{i=1}^{n}(u_i - \mu)^2 = 0. \tag{4.4.7}$$

The first condition yields

$$\mu_L = n^{-1}\sum_{i=1}^{n}u_i, \tag{4.4.8}$$

the second yields

$$\sigma_L^2 = n^{-1}\sum_{i=1}^{n}(u_i - \mu_L)^2. \tag{4.4.9}$$

So μ_L, the maximum likelihood estimate for μ, is just the arithmetic mean, while σ_L^2 is just the sample variance.

Now suppose that the pdf for u is exponential, centered at μ and with scale σ:

$$p(u; \mu, \sigma) = (2\sigma)^{-1}\exp[-\sigma^{-1}|u - \mu|]. \tag{4.4.10}$$

Then

$$L(\mu, \sigma) = (2\sigma)^{-n}\exp\left[-\sigma^{-1}\sum_{i=1}^{n}|u_i - \mu|\right], \tag{4.4.11}$$

$$l(\mu, \sigma) = -n\log(2\sigma) - \sigma^{-1}\sum_{i=1}^{n}|u_i - \mu|. \tag{4.4.12}$$

Hence

$$\frac{\partial l}{\partial \mu} = -\sigma^{-1}\left(\sum_{\mu > u_i}1 - \sum_{\mu < u_i}1\right) = 0, \tag{4.4.13}$$

provided $u_i \neq \mu$ for any i, while

$$\frac{\partial l}{\partial \sigma} = -\frac{n}{\sigma} + \sigma^{-2} \sum_{i=1}^{n} |u_i - \mu| = 0. \tag{4.4.14}$$

So

$$\sigma_L = n^{-1} \sum_{i=1}^{n} |u_i - \mu_L|, \tag{4.4.15}$$

but μ_L is not so easily determined. If n is even, then μ_L should be greater than $n/2$ samples and less than $n/2 + 1$. Let's assume that the samples are ordered:

$$u_1 \leq u_2 \leq \cdots \leq u_{\frac{n}{2}} < u_{\frac{n}{2}+1} \leq \cdots \leq u_n. \tag{4.4.16}$$

Then we should choose μ_L such that

$$u_{\frac{n}{2}} < \mu_L < u_{\frac{n}{2}+1}; \tag{4.4.17}$$

hence

$$-\sum_{i=1}^{n} |u_i - \mu_L| = \sum_{i=1}^{\frac{n}{2}} (u_i - \mu_L) - \sum_{i=\frac{n}{2}+1}^{n} (u_i - \mu_L) \tag{4.4.18}$$

$$= \sum_{i=1}^{\frac{n}{2}} u_i - \sum_{i=\frac{n}{2}+1}^{n} u_i, \tag{4.4.19}$$

which is independent of μ_L! Had we chosen, say

$$u_{\frac{n}{2}-1} < \mu_L < u_{\frac{n}{2}}, \tag{4.4.20}$$

then

$$-\sum_{i=1}^{n} |u_i - \mu_L| = \sum_{i=1}^{\frac{n}{2}-1} (u_i - \mu_L) - \sum_{i=\frac{n}{2}}^{n} (u_i - \mu_L)$$

$$= \sum_{i=1}^{\frac{n}{2}-1} u_i - \sum_{i=\frac{n}{2}}^{n} u_i + 2\mu_L$$

$$= \sum_{i=1}^{\frac{n}{2}} u_i - \sum_{i=\frac{n}{2}+1}^{n} u_i - 2(u_{\frac{n}{2}} - \mu_L). \tag{4.4.21}$$

But the rhs of (4.4.21) is less than the rhs of (4.4.19) by $2(u_{n/2} - \mu_L)$, which is positive by virtue of (4.4.20). So the choice (4.4.17) is maximal. Note that μ_L is only determined within the interval $u_{n/2} < \mu_L < u_{n/2+1}$, and that σ_L is insensitive to the choice. Maximum likelihood estimation is trivial for normal distributions, but less so for others.

Returning to the normal case, notice that we chose μ to maximize l, as in (4.4.5), that is, to minimize

$$\sum_{i=1}^{N} (u_i - \mu)^2 \tag{4.4.22}$$

with respect to μ. The maximum likelihood estimate of μ is the least-squares estimate. Now let's change the perspective slightly. Suppose that $u_1, \ldots, u_n$ are n measurements of a quantity u, and each measurement involves an error $\epsilon_i \equiv u_i - u$ that is independent of the other errors, and distributed as

$$p(\epsilon; v, \theta) = (2\pi\theta^2)^{-1/2} \exp[-(2\theta^2)^{-1}(\epsilon - v)^2], \tag{4.4.23}$$

where the mean v and variance θ^2 are known. That is,

$$p(u_i; v + u, \theta) = (2\pi\theta^2)^{-1/2} \exp[-(2\theta^2)^{-1}(u_i - u - v)^2]. \tag{4.4.24}$$

Exercise 4.4.1

Show that u_L, the maximum likelihood estimate of u, is

$$u_L = n^{-1} \sum_{i=1}^{n} (u_i - v). \tag{4.4.25}$$

Remove the instrument bias, and take the arithmetic mean! Note that

$$E u_L = n^{-1} \sum_{i=1}^{n} E u_i - v$$

$$= n^{-1} \sum_{i=1}^{n} E \epsilon_i + u - v$$

$$= v + u - v$$

$$= u, \tag{4.4.26}$$

where

$$E \epsilon_i \equiv \int_{-\infty}^{\infty} p(\epsilon_i, v, \theta) \epsilon_i \, d\epsilon_i. \tag{4.4.27}$$

The result (4.4.26) tells us that u_L is an unbiased estimate of u.   □

Exercise 4.4.2

Show that

$$E((u_L - u)^2) = \theta^2. \tag{4.4.28}$$

That is, the variance of the error in the maximum likelihood estimate u_L for u is equal to the variance of the measurement errors.   □

Now let's consider randomly erroneous measurements of a random quantity:

$$v = u + \epsilon, \tag{4.4.29}$$

where u is a random variable and ϵ a random measurement error. We shall denote their pdfs as $p(u)$ and $p(\epsilon)$. (This is a poor notation; $p_u(x)$ and $p_\epsilon(x)$ would be better, where $p_u(x)dx$ is the probability that $x < u < x + dx$ and $p_\epsilon(x)dx$ is the probability that $x < \epsilon < x + dx$, but let's try to keep notation simple if imprecise.) The joint pdf for both u AND ϵ is $p(u, \epsilon)$. Then the *marginal* pdfs are

$$p(u) = \int p(u, \epsilon)\, d\epsilon, \quad p(\epsilon) = \int p(u, \epsilon)\, du. \tag{4.4.30}$$

We may also consider the *conditional* distributions $p(u|\epsilon)$ and $p(\epsilon|u)$. The former is the probability distribution of u, given a value for ϵ; the latter is the probability distribution of ϵ, given a value of u. Hence,

$$p(u, \epsilon) = p(u|\epsilon)p(\epsilon) = p(\epsilon|u)p(u). \tag{4.4.31}$$

If u and ϵ are *independent*, then

$$p(u|\epsilon) = p(u), \quad p(\epsilon|u) = p(\epsilon), \tag{4.4.32}$$

and (4.4.31) reduces to the product rule:

$$p(u, \epsilon) = p(u)p(\epsilon). \tag{4.4.33}$$

In the general case, where u and ϵ may be dependent, (4.4.31) becomes *Bayes' Rule* (Cox and Hinkley, 1974):

$$p(u|\epsilon) = p(\epsilon|u)\frac{p(u)}{p(\epsilon)}. \tag{4.4.34}$$

Combining (4.4.30) and (4.4.31) yields

$$p(u) = \int p(u|\epsilon)p(\epsilon)\, d\epsilon, \quad p(\epsilon) = \int p(\epsilon|u)p(u)\, du. \tag{4.4.35}$$

Combining (4.4.34) and (4.4.35) yields

$$p(u) = \int p(\epsilon|u)p(u)\, d\epsilon, \quad p(\epsilon) = \int p(u|\epsilon)p(\epsilon)\, du, \tag{4.4.36}$$

that is,

$$\int p(\epsilon|u)\, d\epsilon = \int p(u|\epsilon)\, du = 1. \tag{4.4.37}$$

Exercise 4.4.3

Examine "Egbert's Table" (see next page) of values for $p(u, \epsilon)$ for a simple case in which $u = 1$ or 2, while $\epsilon = -1, 0$ or 1. The upper number in each of the six boxes is $p(u, \epsilon)$.

		ϵ			
		-1	0	1	$p(u)$
u	1	0.09 0.1	0.72	0.09	
	2	0.05	0.0	0.05	
	$p(\epsilon)$				

(i) Calculate $p(u)$ for each u, and $p(\epsilon)$ for each ϵ. Verify that these distributions are normalized.

(ii) Use (4.4.31) to calculate $p(\epsilon|u)$. Enter the values in the six boxes (a check value is provided in the first box).

(iii) Show that

$$\text{var}\,(\epsilon = 0.28),$$
$$\text{var}\,(\epsilon|u = 2) = 1.0.$$

That is, an "observation" of the random variable u may *increase* the variance of an unknown but dependent random variable ϵ! □

Now suppose that we have n independent data $v_1, \ldots, v_n$. We wish to form the likelihood function $L = \prod_{i=1}^{n} p(v_i)$. We can determine $p(v)$, if we know $p(v|u)$ and $p(u)$. But $p(v|u)$ is the pdf for recording the value v when the true value is u. That is, $p(v|u)$ is the pdf for ϵ, when $\epsilon = v - u$. At this point our sloppy notation fails us, and we must write

$$p(v|u) = p_\epsilon(v - u). \tag{4.4.38}$$

Then

$$p(v) = \int p_\epsilon(v - u)p(u)\,du. \tag{4.4.39}$$

The distributions in the integrand have parameters $E\epsilon, \sigma_\epsilon^2, Eu, \sigma_u^2, \ldots$ which we would like to estimate, using the data $v_1, \ldots, v_n$. We may do so, by solving the maximum likelihood conditions

$$\frac{\partial \ln L}{\partial E\epsilon} = 0, \quad \text{etc.} \tag{4.4.40}$$

Thus we arrive at maximum likelihood estimators, given conditional and marginal distributions.

Exercise 4.4.4

Assume that p_ϵ and p_u are normal. Derive the maximum likelihood estimates of the four parameters, given independent data $v_1, \ldots, v_n$, where $v = u + \epsilon$. □

4.4.3   Bayesian estimation

Let us now apply these ideas to optimal interpolation (Lorenc, 1997). The gridded multivariate field can be ordered as a vector of N components: $\mathbf{u} = (u_1, \ldots, u_N) \in \mathbb{R}^N$.
Assume that we have a prior or "background" estimate $\mathbf{u}^b$, which is usually a model
solution. The prior conditional distribution is $p(\mathbf{u}|\mathbf{u}^b)$. Let there again be M measure-
ments $(v_1, \ldots, v_M) \in \mathbb{R}^M$ related to the true field by $\mathbf{v} = \mathbf{H}(\mathbf{u}) + \epsilon$, where the inde-
pendent error is $\epsilon = (\epsilon_1, \ldots, \epsilon_M) \in \mathbb{R}^M$ and has the pdf $p_\epsilon(\epsilon; E\epsilon, \mathbf{C}_{\epsilon\epsilon}, \ldots)$. Thus $\mathbf{H}$
maps $\mathbb{R}^N$ into $\mathbb{R}^M$. If it is linear, then we may write $\mathbf{H}(\mathbf{u}) = \mathbf{Hu}$, where $\mathbf{H}$ is an $M \times N$
matrix. We want the distribution for $\mathbf{u}$, given the background $\mathbf{u}^b$ and data $\mathbf{v}$. Bayes' Rule
becomes

$$p(\mathbf{u}|\mathbf{v}, \mathbf{u}^b) = \frac{p(\mathbf{v}|\mathbf{u}, \mathbf{u}^b)p(\mathbf{u}|\mathbf{u}^b)}{p(\mathbf{v}|\mathbf{u}^b)}. \tag{4.4.41}$$

Notice that $\mathbf{u}^b$ is being regarded as a fixed parameter here; only $\mathbf{u}$ and $\mathbf{v}$ are being
interchanged. Moreover, the measurement process is unrelated to the model, so

$$p(\mathbf{u}|\mathbf{v}, \mathbf{u}^b) = \frac{p(\mathbf{v}|\mathbf{u})p(\mathbf{u}|\mathbf{u}^b)}{p(\mathbf{v})} \tag{4.4.42}$$

$$\propto p_\epsilon(\mathbf{v} - \mathbf{H}(\mathbf{u}); E\epsilon, \mathbf{C}_{\epsilon\epsilon}, \ldots)p(\mathbf{u}|\mathbf{u}^b). \tag{4.4.43}$$

We may ignore the denominator, as it is independent of $\mathbf{u}$. Given (4.4.43), we take as
our "analysis" estimate of $\mathbf{u}$ the mean value:

$$\mathbf{u}^a = \frac{\int \mathbf{u} \, p(\mathbf{u}|\mathbf{v}, \mathbf{u}^b) \, d\mathbf{u}}{\int p(\mathbf{u}|\mathbf{v}, \mathbf{u}^b) \, d\mathbf{u}}. \tag{4.4.44}$$

Exercise 4.4.5
Suppose that both distributions in (4.4.44) are normal; that is,

$$p_\epsilon(\epsilon; \ldots) = N(\epsilon; E\epsilon, \mathbf{C}_{\epsilon\epsilon}), \tag{4.4.45}$$

$$p(\mathbf{u}|\mathbf{u}^b) = N(\mathbf{u}; \mathbf{u}^b, \mathbf{C}_{uu}). \tag{4.4.46}$$

Derive the standard least-squares optimal interpolation formulae from (4.4.44).   □

   In summary, if we can choose credible distributions for the data error and the background error, be they normal or otherwise, we can use Bayes' Rule to construct a pdf
for the field. Its first moment is a credible estimate of the field.

Exercise 4.4.6
Or is it?   □

4.4.4   Importance sampling

Not all oceanic and atmospheric processes are normally distributed. Not all dynamics and penalty functionals are smooth. There is a need for estimators other than least-squares, and for optimization methods other than the calculus of variations. But first we need a method for synthesizing samples from any probability distribution P. More precisely, we require an algorithm for generating a sequence of real numbers $x_1, x_2, \ldots,$ $x_n, \ldots$ such that the number of values in the interval $(x, x + dx)$ is proportional to $P(x)\,dx$. That is, we wish to perform "importance sampling". It is usually the case that P is in fact a normalized probability distribution function, that is,

$$P(x) = K^{-1}Q(x), \tag{4.4.47}$$

where

$$K = \int_a^b Q(x)\,dx \tag{4.4.48}$$

for some interval $a \le x \le b$. We often only know $Q(x)$ and would like to avoid evaluation of K, especially for higher dimensional problems in which (4.4.48) is a multiple integral.

Consider a Markov chain $x_1, x_2, \ldots, x_n, \ldots,$ for which the value of x_{n+1} lies in the interval $a \le x \le b$, depends only upon the value of x_n and the dependence is random. Let $P_n(x)dx$ be the probability that $x < x_n < x + dx$, and let $T(x, y)dx\,dy$ be the probability that $x < x_{n+1} < x + dx$ given that $y < x_n < y + dy$. Thus, T is a transition probability density, and

$$P_{n+1}(x)\,dx = P_n(x)\,dx + \int_{y=a}^{y=b} \{T(x, y)P_n(y) - T(y, x)P_n(x)\}\,dy\,dx. \tag{4.4.49}$$

The first term in the integrand accounts for transitions to x from all possible y; the second accounts for transitions from x to all possible y. Note that the integral is over y. The chain is in *equilibrium* if $P_{n+1}(x) = P_n(x)$, which implies that the integral in (4.4.49) vanishes. That is the *condition of balance*. Note the assumption that T is independent of n. The *condition of detailed balance*:

$$T(x, y)P(y) - T(y, x)P(x) = 0 \tag{4.4.50}$$

is sufficient but not necessary for equilibrium. Then the Markov chain $x_1, x_2, \ldots$ is in equilibrium, with distribution $P(x)$. A simple algorithm for generating a chain from a given pdf P is as follows (Metropolis *et al.*, 1953).

(1) Pick a number z at random in $[a,b]$.
(2) Calculate

$$r = \frac{P(z)}{P(x_n)}. \tag{4.4.51}$$

(3) Pick a number η at random in $[0,1]$.

(4) Choose

$$x_{n+1} = \begin{cases} z, & \text{if} \quad \eta < r; \\ x_n, & \text{if} \quad \eta > r. \end{cases} \tag{4.4.52}$$

In effect the choice is

$$x_{n+1} = \begin{cases} z, & \text{with probability } r; \\ x_n, & \text{with probability } 1-r. \end{cases} \tag{4.4.53}$$

Hence if $r < 1$ then r is the probability of transition from x_n to z:

$$T(z, x_n) = r = \frac{P(z)}{P(x_n)}, \tag{4.4.54}$$

while if $r > 1$, then

$$T(z, x_n) = 1. \tag{4.4.55}$$

The condition of detailed balance follows immediately. For example, if $r < 1$:

$$T(z, x_n)P(x_n) = \frac{P(z)}{P(x_n)}P(x_n) = P(z) = T(x_n, z)P(z). \tag{4.4.56}$$

Note that the algorithm depends on P only via the ratio (4.4.51). In fact

$$r = \frac{Q(z)}{Q(x_n)}. \tag{4.4.57}$$

It is not necessary to know the normalizing constant (4.4.48).

4.4.5   Substituting algorithms

Suppose that u is the solution of an equation such as

$$\mathcal{D}(u) = f, \tag{4.4.58}$$

where $\mathcal{D}$ is some nonlinear function, while f is a random variable with pdf $A = A(f)$. That is, the probability of $g < f < g + dg$ is $A(g)dg$. What is the corresponding pdf $B = B(u)$? This is a nontrivial analytical problem if $\mathcal{D}$ is nontrivial. However, we can construct a Markov chain $u_1, u_2, \ldots$ distributed according to B. Use importance sampling, based on the non-normalized pdf $Q(u) = A(\mathcal{D}(u))$. For a given u_n, pick z at random, calculate $r = Q(z)/Q(u_n)$, and proceed as in (4.4.52). We would then have properly distributed samples for u, and could form a histogram estimate of its pdf B. These samples are now being loosely described as "ensembles" in the literature. The ensemble is the totality. Note especially that we do not have to invert the operator $\mathcal{D}$. We merely have to evaluate it for each u_n, by direct substitution of u_n into $\mathcal{D}$.

Exercise 4.4.7

Estimate the pdf of u, given that

$$\log_e u = f \qquad (4.4.59)$$

and that f is Gaussian.   □

This approach may be invaluable when f is a random field and $\mathcal{D}$ represents the dynamics of an ocean model. Steady models can be particularly difficult to solve, especially if they are nonlinear. Some time-dependent intermediate models include diagnostic equations that are unwieldy. An obvious example is the stratified quasi-geostrophic model. In particular, diagnosing (solving) the three-dimensional elliptic equation $\nabla^2 \psi = \xi$ for the streamfunction ψ in a realistic ocean basin is nontrivial: assembling sparse matrices requires great care. In comparison, it is relatively trivial to substitute the streamfunction into the elliptic equation, and then substitute the vorticity ξ into the first-order wave equation:

$$\frac{\partial \xi}{\partial t} + J(\psi, \xi + \beta y) = q, \qquad (4.4.60)$$

where β is the local meridional gradient of the Coriolis parameter, and where q is some random source of vorticity.

4.4.6   Multivariate importance sampling

Thus far, u has been a single, real random variable. We are interested in random multivariate fields: $u = u(x, y, z, t)$, $v = v(\cdots)$, $w = w(\cdots)$, $p = p(\cdots)$, etc. In computational practice, these fields are defined on grids, thus we have arrays $u_{ijkl} = u(x_i, y_j, z_k, t_l)$, $v_{ijkl} = v(\cdots)$, $w_{ijkl} = w(\cdots)$, $p_{ijkl} = p(\cdots)$, etc. For clarity, let us condense all these into a single vector $\mathbf{u} = (u_1, \ldots, u_m, \ldots, u_M)$. A Markov chain of these vectors will be denoted by $\mathbf{u}^n = (u_1^n, u_2^n, \ldots, u_M^n)$, for $n = 1, 2, 3, \ldots$. Notice that the upper index n is not the time index; the latter is included in the lower index.

Suppose that the multivariate probability distribution for $\mathbf{u}$ is factorable:

$$Q(\mathbf{u}) = Q_1(u_1)Q_2(u_2)\ldots Q_M(u_M), \qquad (4.4.61)$$

in which case the components of $\mathbf{u}$ are independent. Consider, for example:

$$Q(\mathbf{u}) = \exp\left[-u_1^2 - u_2^2 - \cdots - u_M^2 \right]. \qquad (4.4.62)$$

Then we may apply importance sampling to each component independently. The decision to accept a new value z_m for u_m^{n+1} would be based on the ratio

$$r_m = \frac{Q_m(z_m)}{Q_m\left(u_m^n\right)}. \qquad (4.4.63)$$

These M decisions could be made in series or in parallel. Now suppose that components of $\mathbf{u}$ depend only upon two nearest neighbors:

$$Q(\mathbf{u}) = Q_2(u_1, u_2, u_3)Q_3(u_2, u_3, u_4)Q_4(u_3, u_4, u_5)\ldots Q_{M-1}(u_{M-2}, u_{M-1}, u_M).$$
(4.4.64)

Consider, for example:

$$Q(\mathbf{u}) = \exp[-(u_1 + u_3 - 2u_2)^2 - (u_2 + u_4 - 2u_3)^2$$
$$- \cdots - (u_{M-2} + u_M - 2u_{M-1})^2].$$
(4.4.65)

Then importance sampling may be performed in three "sweeps". Assume that M is divisible by three.

Sweep 1. Choose trial values $z_1, z_4, z_7, \ldots, z_{M-2}$ for $u_1^{n+1}, u_4^{n+1}, u_7^{n+1}, \ldots, u_{M-2}^{n+1}$ respectively; each decision to accept a trial value is independent of the others. For example, the ratio for sampling u_1^{n+1} is

$$r_1 = \frac{Q_2(z_1, u_2^n, u_3^n)}{Q_2(u_1^n, u_2^n, u_3^n)},$$
(4.4.66)

while for u_4^{n+1} it is

$$r_4 = \frac{Q_3(u_2^n, u_3^n, z_4) Q_4(u_3^n, z_4, u_5^n) Q_5(z_4, u_5^n, u_6^n)}{Q_3(u_2^n, u_3^n, u_4^n) Q_4(u_3^n, u_4^n, u_5^n) Q_5(u_4^n, u_5^n, u_6^n)}$$
(4.4.67)

and so on, for $r_7, \ldots, r_{M-2}$. All these decisions can be made in parallel.

Sweep 2. Generate $u_2^{n+1}, u_5^{n+1}, \ldots, u_{M-1}^{n+1}$ by importance sampling, in parallel.

Sweep 3. Generate $u_3^{n+1}, u_6^{n+1}, \ldots, u_M^{n+1}$ by importance sampling, in parallel.

This procedure, known as "checkerboarding", is complicated when the actual computational grid involves more than one dimension.

4.4.7 Simulated annealing

Consider for simplicity a scalar variable u, for which there is a penalty function $J(u)$. Assume only that J is bounded below:

$$J(u) > B$$
(4.4.68)

for all u. We wish to find the value of u for which J is least. Let u_n be an estimate, for which $J_n \equiv J(u_n)$ is unacceptably large. Make a small perturbation to u_n:

$$z = u_n + \Delta u_n.$$
(4.4.69)

The "downhill strategy" is:

$$u_{n+1} = \begin{cases} z, & \text{if} \quad J(z) < J_n \\ u_n, & \text{if} \quad J(z) > J_n. \end{cases}$$
(4.4.70)

However, this strategy could terminate at a local minimum of J. It would be better to allow a few uphill searches at first, in order to avoid such an outcome. So, use importance sampling:

$$u_{n+1} = \begin{cases} z, & \text{if} \quad J(z) < J_n \\ z, & \text{if} \quad J(z) > J_n, \text{ with probability } r \\ u_n, & \text{if} \quad J(z) > J_n, \text{ with probability } 1-r, \end{cases}$$

where

$$r = \frac{e^{-J(z)/\theta}}{e^{-J_n/\theta}} = e^{-(J(z)-J_n)/\theta} < 1 \qquad (4.4.71)$$

for some positive "annealing temperature" θ. Simply pick a random variable η in $[0,1]$. If $\eta < r$, accept z. If $r < \eta$, keep u_n. Now $r \to 0$ as $\theta \to 0$, so fewer uphill steps are allowed as θ decreases. The "annealing strategy", or rate of decrease of θ, is a "black art" (Azencott, 1992). Note that no gradient information for J is used. The penalty function need not even be continuous in u. This approach should be ideal for data assimilation with "small" nonlinear biological models that have "switches". These models typically describe the temporal evolution of a biological system, at one point. The models include constraints such as lower bounds on biomass, together with discontinuous representations of very rapidly adjusting processes. Barth and Wunsch (1990) used simulated annealing to optimize the locations of acoustic transceivers in an idealized model "ocean". Kruger (1993) used simulated annealing to minimize the penalty functionals associated with the inversion of two ocean models. The first was a two-box model of ocean stratification that included a nonsmooth representation of convective adjustment. The second involved a single-level quasigeostrophic model much like (4.4.60), at one time. There were about 4000 computational degrees of freedom in Kruger's second application.

Importance sampling is used extensively in theoretical physics, especially for the evaluation of path integrals. The number of dimensions is extremely large, so the efficient generation of independent trial values is of crucial importance. Ingenious techniques such as "Hybrid Monte Carlo" or HMC, have been devised, but these typically assume that the integrand depends smoothly upon the state variables. See Chapter 6 for an application of these techniques to the resolution of an ill-posed problem.

Chapter 5

The ocean and the atmosphere

Seawater and air are viscous, conducting, compressible fluids. Yet large-scale oceanic and atmospheric circulations have such high Reynolds' numbers and such low aspect ratios that viscous stresses, heat conduction and nonhydrostatic accelerations may all be neglected. (The Mach number of ocean circulation is so low that the compressibility of seawater may also be neglected, but will be retained here in the interest of generality.) Subject to these approximations, the Navier–Stokes equations simplify to the so-called "Primitive Equations". It is often convenient to express these equations in a coordinate system that substitutes pressure for height above or below a reference level. The Primitive Equations were for many years too complex for operational forecasting. They were further reduced by assuming low Rossby number flow, leading to a single equation for the propagation of the vertical component of vorticity – the "quasigeostrophic" equation. Now obsolete as a forecasting tool, this relatively simple equation retains great pedagogical value. To its credit, it is still competitive at predicting the tracks of tropical cyclones, if not their intensity.

The astronomical force that drives the ocean tides is essentially independent of depth, and so its effects may be modeled by unstratified Primitive Equations: the Laplace Tidal Equations. The external Froude number for the tides is so low that the "LTEs" are essentially linear. Combining the linear LTEs with the vast records of sea level elevation collected by satellite altimeters makes an ideal first test for inverse ocean modeling. The interaction of harmonic analysis of the tides and bias-free strategies for measurement leads to novel measurement functionals. The great separation of scales clarifies the prior analysis of errors in the dynamics and in coastlines.

Initializing a quasigeostrophic model for hurricane track prediction is ideal as a first application of inverse ocean modeling to nonlinear dynamics. Errors arise in the dynamics owing to the neglect of resolvable processes. The statistics of these processes may be estimated from archived data. The relatively simple "QG" dynamics also clarify discussion of the conceptual issue of even defining statistics for errors arising from parameterization of unresolvable processes.

High-altitude winds inferred from tracks of cloud images collected by satellites cannot reasonably be assimilated into a "QG" model; a Primitive Equation model is called for. This provides an extreme exercise in deriving and solving Euler–Lagrange equations for a variational principle that is based on a complex model and on an unstructured data set.

The trans-Pacific array of instrumented ocean moorings known as TAO is so regularly structured, and of such continuity and duration, that rigorous testing of models becomes feasible. An intermediate model of the seasonal-to-interannual variability of the coupled ocean–atmosphere is TAO's first victim.

The chapter closes with notes on a selection of contemporary variational and statistical assimilations, variously involving components of the entire hydrosphere.

5.1   The Primitive Equations and the quasigeostrophic equations

5.1.1   Geophysical fluid dynamics is nonlinear

Our development of inverse theory has involved linear models and linear measurement functionals. Tides provide a splendid example of a linear model, but there are no others. In general, geophysical fluid dynamics is nonlinear. The Primitive Equations and the quasigeostrophic equations of motion (Haltiner and Williams, 1980; Gill, 1982) will be briefly reviewed in this section.

5.1.2   Isobaric coordinates

Let us replace space–time coordinates (X, Y, Z, T) with space–pressure–time coordinates (x, y, p, t), where

$$x = X, \quad y = Y, \quad p = p(X, Y, Z, T) \quad \text{and} \quad t = T. \tag{5.1.1}$$

Note 1. We could instead be using, say, spherical coordinates (longitude and latitude) on horizontal surfaces (constant Z), or on isobaric surfaces (constant p) as in (5.1.1).

Note 2. The coordinate transformation (5.1.1) depends upon the state of the ocean or the atmosphere, through the instantaneous pressure field p.

5.1.3 Hydrostatic balance, conservation of mass

In space–time coordinates, the hydrostatic approximation is

$$\frac{\partial p}{\partial Z} = -\rho g, \tag{5.1.2}$$

where ρ is the fluid density, and g the local gravitational acceleration. We may use (5.1.1) and (5.1.2) to obtain the volume element:

$$dx\,dy\,dp = \rho g\,dX\,dY\,dZ. \tag{5.1.3}$$

Now consider a parcel of fluid, occupying a region $V = V(t)$ that moves and distorts in time. The total mass of the parcel does not change, so

$$\frac{d}{dT}\iiint_V \rho\,dX\,dY\,dZ = 0, \tag{5.1.4}$$

or, as a consequence of (5.1.1) and (5.1.3),

$$\frac{d}{dt}\iiint_V dx\,dy\,dp = 0. \tag{5.1.5}$$

Comparing the volume integral at times t and $t + dt$ leads easily to the conclusion that

$$\iint_S \mathbf{v}\cdot\hat{\mathbf{n}}\,da = 0, \tag{5.1.6}$$

where S is the surface of the parcel, $\hat{\mathbf{n}}$ is an outward unit normal on S, da is an element of area in (x, y, p) coordinates, and $\mathbf{v} = (u, v, \omega)$, where

$$u \equiv \frac{Dx}{Dt}, \quad v \equiv \frac{Dy}{Dt}, \quad \omega \equiv \frac{Dp}{Dt}. \tag{5.1.7}$$

The divergence theorem in (x, y, p) coordinates yields

$$\iiint_V \left(\frac{\partial u}{\partial x} + \frac{\partial v}{\partial y} + \frac{\partial \omega}{\partial p}\right) dx\,dy\,dp = 0. \tag{5.1.8}$$

The parcel V is arbitrary, hence the flow is volume-conserving in (x, y, p) coordinates:

$$\boxed{\frac{\partial u}{\partial x} + \frac{\partial v}{\partial y} + \frac{\partial \omega}{\partial p} = 0} \tag{5.1.9}$$

5.1.4   Pressure gradients

The pressure gradient per unit mass is

$$\rho^{-1} \frac{\partial p}{\partial X} = -g \left(\frac{\partial p}{\partial Z} \right)^{-1} \frac{\partial p}{\partial X}, \qquad (5.1.10)$$

by virtue of the hydrostatic approximation (5.1.2). The chain rule applied to $Z = Z(x, y, p, t)$ yields

$$1 = \frac{\partial Z}{\partial Z} = \frac{\partial Z}{\partial x} \frac{\partial x}{\partial Z} + \frac{\partial Z}{\partial y} \frac{\partial y}{\partial Z} + \frac{\partial Z}{\partial p} \frac{\partial p}{\partial Z} + \frac{\partial Z}{\partial t} \frac{\partial t}{\partial Z}. \qquad (5.1.11)$$

But $x = X$, $y = Y$ and $t = T$ are orthogonal to Z, hence

$$\frac{\partial x}{\partial Z} = \frac{\partial y}{\partial Z} = \frac{\partial t}{\partial Z} = 0, \qquad (5.1.12)$$

and so

$$\left(\frac{\partial p}{\partial Z} \right)^{-1} = \left(\frac{\partial Z}{\partial p} \right). \qquad (5.1.13)$$

Note that (5.1.13) is not a general property of transformations; it is only true for our special transformation (5.1.1). The hydrostatic approximation then becomes

$$\boxed{\frac{\partial \phi}{\partial p} = -\rho^{-1}} \qquad (5.1.14)$$

where we have defined the geopotential $\phi = \phi(x, y, p, t)$:

$$\phi \equiv gZ. \qquad (5.1.15)$$

Combining (5.1.10) and (5.1.13) yields

$$\rho^{-1} \frac{\partial p}{\partial X} = -g \frac{\partial Z}{\partial p} \frac{\partial p}{\partial X}. \qquad (5.1.16)$$

Next we use the chain rule on $Z = Z(x, y, p, t)$ to obtain

$$0 = \frac{\partial Z}{\partial X} = \frac{\partial Z}{\partial x} \frac{\partial x}{\partial X} + \frac{\partial Z}{\partial y} \frac{\partial y}{\partial X} + \frac{\partial Z}{\partial p} \frac{\partial p}{\partial X} + \frac{\partial Z}{\partial t} \frac{\partial t}{\partial X}. \qquad (5.1.17)$$

But $\frac{\partial x}{\partial X} = 1$, $\frac{\partial y}{\partial X} = 0$ and $\frac{\partial t}{\partial X} = 0$, so (5.1.10) becomes

$$\rho^{-1} \frac{\partial p}{\partial X} = g \frac{\partial Z}{\partial x} = \frac{\partial \phi}{\partial x}. \qquad (5.1.18)$$

Similarly,

$$\rho^{-1}\frac{\partial p}{\partial Y} = \frac{\partial \phi}{\partial y}.$$ (5.1.19)

Exercise 5.1.1
Draw a sketch that explains (5.1.18) and (5.1.19). □

5.1.5 Conservation of momentum

Now $\mathbf{x} = \mathbf{X}$ and $t = T$, so

$$\mathbf{U} \equiv \frac{D\mathbf{X}}{DT} = \frac{D\mathbf{x}}{Dt} \equiv \mathbf{u}, \quad \frac{D\mathbf{U}}{DT} = \frac{D^2\mathbf{X}}{DT^2} = \frac{D^2\mathbf{x}}{Dt^2} = \frac{D\mathbf{u}}{Dt},$$ (5.1.20)

where $\mathbf{u} = (u, v)$. Thus the horizontal momentum equation

$$\frac{D\mathbf{U}}{DT} + f\hat{\mathbf{k}} \times \mathbf{U} = -\rho^{-1}\nabla_X p,$$ (5.1.21)

where $\nabla_X = \left(\frac{\partial}{\partial X}, \frac{\partial}{\partial Y}\right)$ and $f = f(Y)$ is the (known) Coriolis parameter, becomes the isobaric momentum equation

$$\boxed{\frac{D\mathbf{u}}{Dt} + f\hat{\mathbf{k}} \times \mathbf{u} = -\nabla_x \phi}$$ (5.1.22)

where

$$\frac{D}{Dt} = \frac{\partial}{\partial t} + \mathbf{u} \cdot \nabla_x + \omega\frac{\partial}{\partial p},$$ (5.1.23)

and $\nabla_x = \left(\frac{\partial}{\partial x}, \frac{\partial}{\partial y}\right)$.

5.1.6 Conservation of scalars

For any conserved tracer τ such as entropy η, salinity S or relative humidity q,

$$\frac{D\tau}{DT} = 0.$$

But $T = t$, so

$$\boxed{\frac{D\tau}{Dt} = 0}$$ (5.1.24)

We have now derived the *Primitive Equation* in pressure coordinates: (5.1.9), (5.1.14), (5.1.22) and (5.1.24). Note that only the hydrostatic approximation has been made;

incompressibility has *not* been assumed. The appearance of (5.1.9) is deceptive! The Primitive Equations must be supplemented with an equation of state such as

$$\rho = \rho(p, \eta, S) \qquad (5.1.25)$$

or

$$\rho = \rho(p, \eta, q) \qquad (5.1.26)$$

where the rhs of (5.1.25), (5.1.26) indicates a prescribed functional form.

Exercise 5.1.2
Consider hydrostatic motion of a fluid of constant density, between a rigid flat surface at $\phi = 0$ and a material free surface at $\phi = gh$, where $h = h(x, y, t)$. Assume that the pressure at the free surface vanishes: $p = 0$. Derive the "shallow water equations":

$$\frac{D\mathbf{u}}{Dt} + f\hat{\mathbf{k}} \times \mathbf{u} = -g\nabla h, \qquad (5.1.27)$$

$$\frac{Dh}{Dt} + h\nabla \cdot \mathbf{u} = 0. \qquad (5.1.28)$$

Note: It is dynamically consistent to assume that $\partial\mathbf{u}/\partial p = 0$. □

5.1.7 Quasigeostrophy

The Eulerian form of (5.1.22) is

$$\frac{\partial\mathbf{u}}{\partial t} + (\mathbf{u} \cdot \nabla)\mathbf{u} + \omega\frac{\partial\mathbf{u}}{\partial p} + f\hat{\mathbf{k}} \times \mathbf{u} = -\nabla\phi, \qquad (5.1.29)$$

where it is now understood that ∇ is ∇_x. For mesoscale motions and larger, it is reasonable to assume that

$$\left|\frac{\partial\omega}{\partial p}\right| \ll \left|\nabla \cdot \mathbf{u}\right|, \qquad \left|\omega\frac{\partial\mathbf{u}}{\partial p}\right| \ll \left|\mathbf{u} \cdot \nabla\mathbf{u}\right|, \qquad (5.1.30)$$

hence the Primitive Equations (5.1.9) and (5.1.22) are approximately

$$\nabla \cdot \mathbf{u} = 0, \qquad (5.1.31)$$

$$\frac{\partial\mathbf{u}}{\partial t} + (\mathbf{u} \cdot \nabla)\mathbf{u} + f\hat{\mathbf{k}} \times \mathbf{u} = -\nabla\phi, \qquad (5.1.32)$$

which is the same as the dynamics of planar incompressible flow. Note that ∇ is a gradient at constant pressure, and that (5.1.31) and (5.1.32) form a closed system, to the extent that they determine $\mathbf{u}$ and ϕ without reference to the equation of state, or to the hydrostatic approximation or to the conservation of entropy. As is well known, in simply-connected domains (5.1.31) implies the existence of a streamfunction $\psi = \psi(\mathbf{x}, p, t)$ such that

$$\mathbf{u} = \hat{\mathbf{k}} \times \nabla\psi = \left(-\frac{\partial\psi}{\partial y}, \frac{\partial\psi}{\partial x}\right). \tag{5.1.33}$$

If we define the vertical component of relative vorticity by

$$\xi \equiv \hat{\mathbf{k}} \cdot \nabla \times \mathbf{u} = \frac{\partial v}{\partial x} - \frac{\partial u}{\partial y} = \nabla^2\psi, \tag{5.1.34}$$

and if we apply $\hat{\mathbf{k}} \cdot \nabla \times$ to (5.1.32) and use (5.1.31), we obtain the vorticity equation

$$\frac{\partial\xi}{\partial t} + \mathbf{u} \cdot \nabla\xi + \frac{df}{dy}v = 0, \tag{5.1.35}$$

or

$$\boxed{\frac{\partial\zeta}{\partial t} + \mathbf{u} \cdot \nabla\zeta = 0} \tag{5.1.36}$$

where

$$\zeta = \xi + f \tag{5.1.37}$$

is the total vorticity. The conservation law (5.1.36) may be expressed entirely in terms of the streamfunction:

$$\boxed{\frac{\partial}{\partial t}\nabla^2\psi + \frac{\partial(\psi, \nabla^2\psi + f)}{\partial(x, y)} = 0} \tag{5.1.38}$$

The nonlinearity of this "filtered" vorticity equation is significant for large Rossby number:

$$Ro_\beta \equiv \frac{U}{\beta l^2} \gtrsim 1, \tag{5.1.39}$$

where $\beta \equiv \frac{df}{dy}$, while U and l are the scales of variation of $\mathbf{u}$ and $\mathbf{x}$ respectively. Note that (5.1.39) may still hold even though

$$Ro_f \equiv \frac{U}{f_0 l} \ll 1, \quad \beta l \ll f_0, \tag{5.1.40}$$

where f_0 is a local value of f. Under these conditions, (5.1.29) yields the "geostrophic" balance:

$$f_0\hat{\mathbf{k}} \times \mathbf{u} \cong -\nabla\phi, \tag{5.1.41}$$

in which case (5.1.31) again holds approximately, and (5.1.35), (5.1.36) may be derived from (5.1.29) at $O(Ro_f)$. As a consequence, (5.1.36) is also known as the "quasigeostrophic" vorticity equation (Gill, 1982).

5.2   Ocean tides

5.2.1   Altimetry

Barotropic ocean tides are global-scale motions that are accurately modeled with linear dynamics. The TOPEX/POSEIDON altimetric satellite, launched in August 1992 and still in operation (June, 2001), is providing highly accurate global sea-level data (Chelton *et al.*, 2001). There could hardly be a more elegant exercise in data assimilation. Indeed, altimetry provided a first test of an advanced method using a large amount of data. Let us pause in our development of inverse methods, and explore the real problem of global ocean tides.

5.2.2   Lunar tides

Let us briefly review lunar tides. They are driven by the gravitational attraction of the moon: see Fig. 5.2.1. At the earth's center of mass E, there is exact equality between the gravitational attraction towards the center of mass of the moon at C and the centripetal acceleration of E towards C, as E moves tangentially on its orbit around C. (More precisely, C and E orbit around the common center of mass.) Let the points A and B make orbits of the same radius as that of E, and so have the same centripetal acceleration as E. However, A is closer to C than is E (while B is further away), and so A experiences a stronger gravitational acceleration towards C (while B experiences a

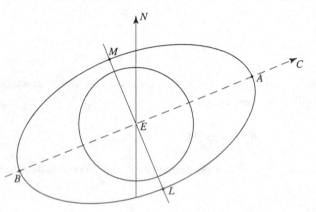

Figure 5.2.1 Tidal potentials.

weaker gravitational acceleration). Hence there are net accelerations or "tidal bulges" at A and B, respectively towards and away from C. Meanwhile the earth spins around its polar axis once a day, by definition of the polar axis and the day. So each point on the ocean surface should have two high tides (A and B) and two low tides (L and M) each day. In fact, the ocean has free barotropic motions with many periods of the order of a day, hence its response to the tide-generating force is very complicated. Solar tides add to the complexity. For a marvelous account of tides in the ocean, see Cartwright (1999).

The tide-generating force (tgf) is conservative, and so there is a tide-generating potential (tgp) per unit mass which we shall express as a sea-surface elevation $\overline{h}$. The tgf is $\nabla \overline{h}$. In mid-ocean, $|\overline{h}|$ is about 30 cm. The tgf has a complicated time dependence, dominated by the relative motions of the earth, the moon and the sun. Certain periodicities are obvious, and up to 400 others have been calculated using celestial mechanics, by G. Darwin, Doodson and others. That is,

$$\overline{h}(\mathbf{x}, t) \cong \mathcal{R}e \sum_{k=1}^{K} \overline{h}_k(\mathbf{x})e^{i\omega_k t}, \tag{5.2.1}$$

where $\mathbf{x}$ denotes a position on the earth's surface. The frequencies $\omega_1, \ldots, \omega_K$ define tidal constituents. For example $\omega_1 \equiv$ 'M_2', the "principal lunar semidiurnal constituent", corresponds to a period of 12h 25m 42s approx., while $\omega_2 \equiv$ 'S_2', the "principal solar semidiurnal constituent", corresponds to a period of 12h exactly. Table 5.2.1 is extracted from Doodson and Warburg (1941). The "speed number" is the frequency ω_k expressed in degrees per hour (and is equal to exactly 30 for S_2), while the "relative coefficient" is the relative amplitude of $\overline{h}_k$. The dominant diurnal and semidiurnal constituents are, in order, M_2, K_1, S_2, O_1, P_1, N_2, K_2 and Q_1. The lunar fortnightly, monthly and solar semiannual tides Mf, Mm and Ssa are also significant.

Constructive and destructive interference between semidiurnal and diurnal tides causes a diurnal inequality, that is, one of the two daily high tides exceeds the other. Interference between semidiurnal tides, especially between M_2 and S_2, causes beating or "neap" and "spring" tides.

Exercise 5.2.1
What is the period of beats between M_2 and S_2?   □

5.2.3   Laplace Tidal Equations

Having briefly reviewed the tide-generating force, let us now review ocean hydrodynamics. It suffices to consider the linear, shallow-water equations on a rotating planet

Table 5.2.1 List of harmonic constituents of the equilibrium tide on the Greenwich Meridian

Symbol	Argument	Speed number	Relative coefficient
Sa	h	0.0411	0.012
Ssa	$2h$	0.0821	0.073
Mm	$s - p$	0.5444	0.083
MSf	$2s - 2h$	1.0159	0.014
Mf	$2s$	1.0980	0.156
K_1	$15°t + h + 90°$	15.0411	0.531
O_1	$15°t + h - 2s - 90°$	13.9430	0.377
P_1	$15°t - h - 90°$	14.9589	0.176
Q_1	$15°t + h - 3s + p - 90°$	13.3987	0.072
$\tilde{M}_1$	$15°t + h - s + 90°$	14.4921	0.040
J_1	$15°t + h + s - p + 90°$	15.5854	0.030
M_2	$30°t + 2h - 2s$	28.9841	0.908
S_2	$30°t$	30.0000	0.423
N_2	$30°t + 2h - 3s + p$	28.4397	0.174
K_2	$30°t + 2h$	30.0821	0.115
ν_2	$30°t + 4h - 3s - p$	28.5126	0.033
μ_2	$30°t + 4h - 4s$	27.9682	0.028
L_2	$30°t + 2h - s - p + 180°$	29.5285	0.026
T_2	$30°t - h + p'$	29.9589	0.025
$2N_2$	$30°t + 2h - 4s + 2p$	27.8954	0.023

(the Laplace Tidal Equations or LTEs). In the f-plane approximation (Gill, 1982), these are

$$\frac{\partial \mathbf{u}}{\partial t} + f\hat{\mathbf{k}} \times \mathbf{u} = -g\nabla(h - \overline{h}) - r\mathbf{u}/H, \tag{5.2.2}$$

$$\frac{\partial h}{\partial t} + \nabla \cdot (H\mathbf{u}) = 0, \tag{5.2.3}$$

where f is the local value of the Coriolis parameter, $\hat{\mathbf{k}}$ is the unit vector in the local vertically-upward direction, $\mathbf{u} = \mathbf{u}(\mathbf{x}, t)$ is the barotropic current, $h = h(\mathbf{x}, t)$ is the sea-level disturbance, $H = H(\mathbf{x})$ is the mean depth of the ocean, r is a bottom drag coefficient and $\overline{h} = \overline{h}(\mathbf{x}, t)$ is the tgp: see Fig. 5.2.2.

Note 1. A quadratic drag law $-k|\mathbf{u}|\mathbf{u}$ is more reliable.

Note 2. If $h \equiv \overline{h}$, then the ocean is in hydrostatic balance with the tgf: this is the "equilibrium" tide of Newton. For long-period tides such as Mf, Mm and Ssa, it is an excellent approximation.

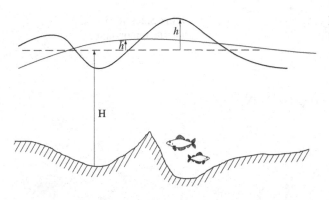

Figure 5.2.2 Shallow-water theory.

Note 3. Many more effects can be included, yielding real gains in forecast accuracy. These include

- (i) load tide: as sea level rises and falls, the ocean floor subsides and rebounds elastically;
- (ii) earth tides: the tgf directly drives motions in the elastic earth;
- (iii) self tide: as sea levels rises, the local accumulation of mass deflects the local vertical;
- (iv) geoid corrections: the earth tides change the shape of the earth and hence that of the earth's geopotentials or "horizontals";
- (v) atmospheric tides: the tgf and solar heating drive motions in the atmosphere which perturb sea-level pressure.

The LTEs require boundary conditions, such as

$$\mathbf{u} \cdot \hat{\mathbf{n}} = 0 \qquad (5.2.4)$$

at coasts, or

$$h = h_B \qquad (5.2.5)$$

at an open boundary. These are unsatisfactory: is the boundary at the shore line or the shelf break? Can h_B be measured economically? How shall we avoid spurious oscillations in an open region, when the LTEs are subjected to (5.2.5)?

5.2.4   Tidal data

Tides are the best measured of all ocean phenomena. The data include:

- (i) century-long high-quality time series of sea level at about one hundred coastal stations, measured with floats in "stilling wells" and strip-chart recorders;
- (ii) year-long high-quality time series of bottom pressure in about twenty deep ocean locations, measured with the piezoelectric effect and digital recorders;
- (iii) year-long good-quality time series of ocean current at selected depths at about a thousand deep locations;

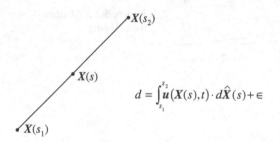

Figure 5.2.3 Reciprocal-shooting tomography.

$$d = \int_{s_1}^{s_2} \boldsymbol{u}(X(s),t) \cdot d\hat{X}(s) + \epsilon$$

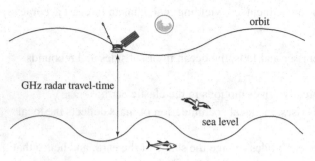

Figure 5.2.4 Satellite radar altimetry.

(iv) year-long good-quality time series of reciprocal-shooting acoustic tomography at about a dozen deep ocean locations: see Fig. 5.2.3.

(v) satellite altimetry: see Fig. 5.2.4.

Altimetric missions include GEOSAT, ERS-1 and TOPEX/POSEIDON. The last (T/P) has a ten-day repeat-track orbit from 70°S to 70°N approx.: see Fig. 5.2.5. Again, TOPEX/POSEIDON has been flying and operating successfully since August 1992.

Temporal variability in the orbit of T/P is known with remarkable precision: ±2 cm. However, the shape of the gravitational equipotential (the "geoid") is not known so accurately. This bias can be eliminated from the data by considering "cross-over" differences: see Fig. 5.2.6. The datum becomes

$$d = h(\mathbf{X}, T_D) - h(\mathbf{X}, T_A) + \epsilon.$$

Note that T_A and T_D need not be the times of *consecutive* passes over $\mathbf{X}$. The values of $|T_A - T_D|$ for consecutive passes can be as large as five days, thus semi-diurnal tides are severely aliased. In fact, the aliased tides resemble Rossby waves with periods of about 60 days. We shall use the dynamics of the LTE to identify and hence reject the aliased tides, which have great spatial coherence.

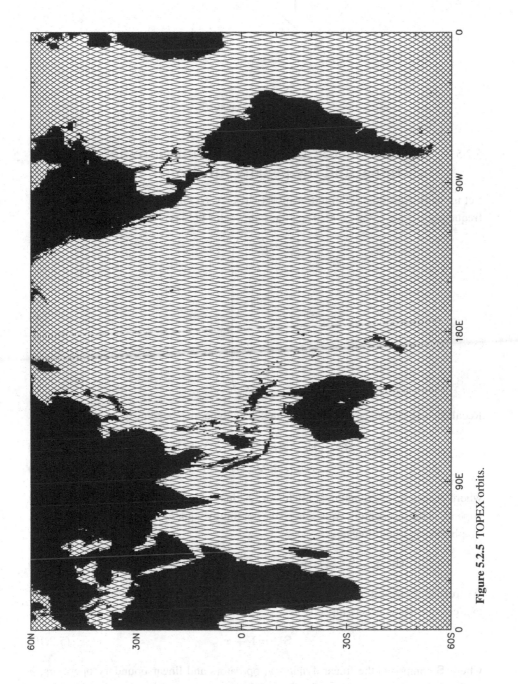

Figure 5.2.5 TOPEX orbits.

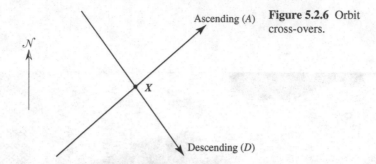

Ascending (A)

Figure 5.2.6 Orbit cross-overs.

Descending (D)

5.2.5 The state vector, the cross-over measurement functional and the penalty functional

Let us work with tidal volume transports $U_k(\mathbf{x}) \equiv H(\mathbf{x})\mathbf{u}_k(\mathbf{x})$ and elevations $h_k(\mathbf{x})$ at frequency ω_k, for $1 \le k \le K$. Then the state vector field is

$$
\mathbf{U} = \mathbf{U}(\mathbf{x}) = \begin{pmatrix} \mathbf{U}_1 \\ h_1 \\ \mathbf{U}_2 \\ h_2 \\ \vdots \\ \vdots \\ \mathbf{U}_K \\ h_K \end{pmatrix} \in \mathbb{C}^{3K}. \tag{5.2.6}
$$

Recall that $\mathbf{U}_k$ and h_k are complex-valued. The dynamics become

$$
i\omega_k \mathbf{U}_k + f\hat{\mathbf{k}} \times \mathbf{U}_k + gH\boldsymbol{\nabla}(h_k - \overline{h}_k) + r\mathbf{U}_k/H = \rho_k, \tag{5.2.7}
$$

$$
i\omega_k h_k + \boldsymbol{\nabla} \cdot \mathbf{U}_k = \sigma_k, \tag{5.2.8}
$$

where ρ_k and σ_k are dynamical misfits or residuals. Note that the bathymetry H only appears in the momentum equation, so the continuity equation should be very accurate. The boundary conditions are

$$
\mathbf{U}_k \cdot \hat{\mathbf{n}} \cong 0 \quad \text{at coasts,} \quad h_k \cong h_{B_k} \quad \text{at open boundaries.} \tag{5.2.9}
$$

We refrain from introducing symbols for the boundary residuals; instead we shall express (5.2.7)–(5.2.9) compactly as

$$
\mathbf{SU} = \mathbf{F} + \tau, \tag{5.2.10}
$$

where $\mathbf{S}$ comprises the linear dynamical operators and linear boundary operators, $\mathbf{F}$ includes the tgf and boundary forcing, while τ includes the dynamical and boundary residuals.

The cross-over data involve the linear measurement functional

$$\mathcal{L}_m[\mathbf{U}] = h\left(\mathbf{X}_m, T_m^{(2)}\right) - h\left(\mathbf{X}_m, T_m^{(1)}\right),$$ (5.2.11)

where $\mathbf{X}_m$ is a cross-over location, while $T_m^{(1)}$ and $T_m^{(2)}$ are the times of distinct passes. Note that $\mathcal{L}_m$ selects only the elevation h, and evaluates it at certain times. In terms of complex harmonic amplitudes, the functional becomes

$$\mathcal{L}_m[\mathbf{U}] = \Re e \sum_{k=1}^{K} h_k(\mathbf{X}_m)\left\{e^{i\omega_k T_m^{(2)}} - e^{i\omega_k T_m^{(1)}}\right\},$$ (5.2.12)

for $1 \le m \le M$. In compact form, the data are

$$\mathbf{d} = \mathcal{L}[\mathbf{U}] + \boldsymbol{\epsilon}.$$ (5.2.13)

Note that the field $\mathbf{U}$ has values in $\mathbb{C}^{3K}$, while $\mathbf{d}$ belongs to $\mathbb{R}^M$.

Exercise 5.2.2
Devise the measurement functionals for data types (i)–(iv) in §5.2.3. Explain why representers for altimetric cross-over data can be constructed using representers for tide gauge data. $\square$

Finally, the penalty functional for inversion of the LTE and T/P data is (Egbert et al., 1994; Egbert and Bennett, 1996)

$$\mathcal{J}[\mathbf{U}] = \boldsymbol{\tau}^* \circ \mathbf{W}_\tau \circ \boldsymbol{\tau} + \boldsymbol{\epsilon}^* \mathbf{w}\boldsymbol{\epsilon},$$ (5.2.14)

where * denotes the transposed complex conjugate vector. Note that we should choose $\mathbf{W}_\tau = \mathbf{C}_\tau^{-1}$, where $\mathbf{C}_\tau$ is the covariance of residuals at different places and at different frequencies, and we should choose $\mathbf{w} = \mathbf{C}_\epsilon^{-1}$ where $\mathbf{C}_\epsilon$ is the covariance of measurement errors.

5.2.6   Choosing weights: scale analysis of dynamical errors

Before proceeding with the mathematical task of minimizing the penalty functional $\mathcal{J}$, let us take a first look at the choice of weights in $\mathcal{J}$. As these will be the inverses of error covariances, consider first some scale estimates of dynamical errors.

(i) The dynamics are linearized. Also, we have analyzed the fields harmonically, thus $\frac{\partial h}{\partial t} = i\omega h$ at frequency ω. Let us assume that $\frac{\partial h}{\partial x} \sim \kappa \delta h$, where κ is some wavenumber and δh is the rough magnitude of h. The balance between local accelerations and pressure gradients (5.2.7) may be expressed as

$$\omega \delta u \sim g\kappa \delta h.$$ (5.2.15)

The balance between the local rate of change of sea level and the convergence of volume flux in (5.2.8) is

$$\omega \delta h \sim \kappa H \delta u. \tag{5.2.16}$$

Hence

$$\omega^2 \sim g H \kappa^2, \quad \kappa \sim \frac{\omega}{c}, \tag{5.2.17}$$

where $c = (gH)^{\frac{1}{2}}$, and

$$\delta u \sim \frac{\omega \delta h}{\kappa H} \sim \left(\frac{g}{H}\right)^{\frac{1}{2}} \delta h = c \frac{\delta h}{H}. \tag{5.2.18}$$

Now compare the local acceleration and momentum advection in (5.2.7):

$$\omega \delta u : \kappa (\delta u)^2. \tag{5.2.19}$$

These are in the ratio

$$1 : \frac{\kappa \delta u}{\omega}, \quad \text{or} \quad 1 : \frac{\delta u}{c}, \quad \text{or} \quad 1 : \frac{\delta h}{H}. \tag{5.2.20}$$

The linearization error in the continuity equation (5.2.8) is also $(\delta h/H)$. In deep water $H \sim 5000$ m and $\delta h \sim 0.2$ m, so linearization is highly accurate. Note that $c = (gH)^{\frac{1}{2}} \sim 200 \text{ m s}^{-1}$, so $\delta u \sim 0.008 \text{ m s}^{-1}$.

(ii) The pressure gradients in (5.2.7) are derived from the hydrostatic balance (not shown). Using the three-dimensional incompressibility condition (not shown), we may deduce that the scale of the vertical velocity is $\delta w \sim \kappa H \delta u$, hence the comparison of local vertical accelerations to the gravitational acceleration is

$$\omega \delta w : g, \quad \text{or} \quad \omega \kappa H \delta u : g, \tag{5.2.21}$$

or

$$c \kappa^2 H \delta u : g, \quad \text{or} \quad \kappa^2 H^2 (\delta u / c) : 1. \tag{5.2.22}$$

So the hydrostatic balance is extremely accurate for small-amplitude ($\delta h \ll H$), long waves ($\kappa H \ll 1$) in deep water. The dynamics are "shallow" in the sense that $\kappa H \ll 1$.

(iii) A crude estimate of numerical accuracy is made by comparing the horizontal grid spacing Δx to the length scale $\kappa^{-1} = c\omega^{-1}$. For solar semidiurnal tides,

$$\omega = 2\pi \left(\frac{1}{2}d\right)^{-1} = (2\pi/43\,200) \text{ s}^{-1} \cong 1.4 \times 10^{-4} \text{ s}^{-1},$$

so $\kappa^{-1} \sim 200 \text{ m s}^{-1}/(1.4 \times 10^{-4} \text{ s}^{-1}) \cong 1.4 \times 10^6$ m $= 1400$ km. Thus, if $\Delta x = 0.5° \cong 50$ km, and the numerics are second-order accurate, then truncation errors are entirely negligible. Tidal diffraction at peninsulae reduces the length scale significantly. It is also common practice to reduce grid spacing in shallow water according to the rule $\Delta x \propto H^{\frac{1}{2}}$.

Exercise 5.2.3

Justify the shallow-water grid rule given above.   □

  (iv) The mean depth $H(\mathbf{x})$ is commonly taken from the US Navy's ETOP95 bathymetry, which is available at NCAR. These data are very doubtful at high latitudes. In the deep North Pacific we can only guess that the error is about 100 m in 5000 m, or 2%. There are known to be far greater errors in, for example, the Weddell Sea.

  (v) We have adopted the crude drag law: $i\omega\mathbf{u} \cdots = \cdots - r\mathbf{u}/H$, where $r = 0.03$ m s^{-1}. It is common practice to replace r/H with $r/\max[H, 200\,\mathrm{m}]$, in order to avoid excessive drag over the continental shelves. These drag formulae are usually tuned so that the tidal solutions are in reasonable agreement with data. Nevertheless, such drag laws are crude parameterizations, so it is prudent to assume that they are 100% in error. However, the drag is a very small part of the momentum balance in deep water.

  (vi) The rigid boundary condition is simply

$$H\mathbf{u} \cdot \hat{\mathbf{n}} = 0, \tag{5.2.23}$$

where $\hat{\mathbf{n}}$ is normal to the boundary. The question arises: where is the boundary? In a numerical model the precision of location is no smaller than Δx, so the error in (5.2.23) is of the order of

$$\Delta x \frac{\partial}{\partial n}(H\mathbf{u} \cdot \hat{\mathbf{n}}) \sim \Delta x \kappa H \delta \mathbf{u} \cdot \hat{\mathbf{n}}. \tag{5.2.24}$$

The relative error in (5.2.23) is therefore $\sim \Delta x \kappa$. If we assume that $\kappa \sim \omega(gH)^{-\frac{1}{2}}$, $H \sim 100$ m, $g \sim 10$ m s^{-2} and $\omega = S_2 \simeq 1.4 \times 10^{-4}$ s^{-1}, then $\kappa \cong 0.5 \times 10^{-5}$ m^{-1}. So if $\Delta x = 0.5° \cong 50$ km, then

$$\Delta x \kappa \cong 0.25. \tag{5.2.25}$$

The relative error in (5.2.23) is 25%! The depth would have to increase to 10 km in order for $\Delta x \kappa$ to be as small as 2.5% (given $\Delta x = 0.5°$). So rigid boundary conditions are significant sources of error in numerical tidal models. The solution in mid-ocean may not be sensitive to this error source, as the basin resonances are very broad. That is, the coastal irregularity itself ensures a fine spectrum of seiche modes. Finally, the high-resolution Finite Element Model (FEM) for global tides developed at the Institute for Mechanics in Grenoble, France is the best forward model yet developed (Le Provost et al., 1994).

In summary, linearization and truncation errors in the continuity equation are negligible. Bathymetric errors and drag errors in the momentum-transport equations should be admitted, while rigid boundary conditions are significantly in error.

5.2.7   The formalities of minimization

Let us set aside our preliminary discussion of model errors, and make some notes on the formalities of minimization. The penalty functional (5.2.14) is

$$\mathcal{J}[U] = \tau^* \circ \mathbf{C}_\tau^{-1} \circ \tau + \epsilon^* \mathbf{C}_\epsilon^{-1} \epsilon \qquad (5.2.26)$$

$$\equiv (\mathbf{SU} - \mathbf{F})^* \circ \mathbf{C}_\tau^{-1} \circ (\mathbf{SU} - \mathbf{F}) + (\mathbf{d} - \mathcal{L}[U])^* \mathbf{C}_\epsilon^{-1} (\mathbf{d} - \mathcal{L}[U]). \qquad (5.2.27)$$

Setting the first variation of $\mathcal{J}$ to zero yields

$$0 = \frac{1}{2}\delta\mathcal{J} = (\mathbf{S}\delta U)^* \circ \mathbf{C}_\tau^{-1} \circ (\mathbf{S}\hat{U} - \mathbf{F}) - \mathcal{L}[\delta U]^* \mathbf{C}_\epsilon^{-1} (\mathbf{d} - \mathcal{L}[\hat{U}]). \qquad (5.2.28)$$

The vanishing of the coefficient of δU^* yields

$$\mathbf{S}^\dagger \Lambda = \mathcal{L}[\delta]^* \mathbf{C}_\epsilon^{-1} (\mathbf{d} - \mathcal{L}[\hat{U}]), \qquad (5.2.29)$$

where

$$\mathbf{S}\hat{U} = \mathbf{F} + \mathbf{C}_\tau \circ \Lambda. \qquad (5.2.30)$$

Note that $\mathbf{S}$ and the *adjoint operator* $\mathbf{S}^\dagger$ include the dynamics and the boundary conditions.

> **Exercise 5.2.4**
> Derive (5.2.29), (5.2.30) in detail.   □

Let us now examine $\mathcal{L}$ for TOPEX/POSEIDON cross-over data (T/P XO data):

$$\mathcal{L}_m[U] = h(\mathbf{X}_i, T_j^{(2)}) - h(\mathbf{X}_i, T_j^{(1)}), \qquad (5.2.31)$$

where $1 \leq m = m(i, j) \leq M$. The $\mathbf{X}_i$ for $1 \leq i \leq I$ are the XO locations; the $T_j^{(1,2)}$ for $1 \leq j \leq J$ are the XO times. In terms of tidal constituents we have, from (5.2.12):

$$\mathcal{L}_m[U] = \Re \sum_{k=1}^{K} h_k(\mathbf{X}_i)\{e^{i\omega_k T_j^{(2)}} - e^{i\omega_k T_j^{(1)}}\}. \qquad (5.2.32)$$

So it suffices to calculate representers for $h_k(\mathbf{X}_i)$ for $1 \leq i \leq I$ and $1 \leq k \leq K$. Then we can synthesize the representers for the $(\mathbf{X}_i, T_j^{(1,2)})$ XO difference. This is very useful. There are only 1×10^4 XO points but by 9/99 there had been approximately 258 ten-day repeat-track orbit cycles, or about 1.8×10^6 XO data. According to the above harmonic analysis, we need only compute $K \times 10^4$ representers (K is usually 4 or 8). How else might we reduce the computations? Inspection of reasonably accurate solutions of forward tidal models indicates that the XO coverage is unnecessarily dense, for observing tides. In the open ocean, adequate coverage is obtained with every third XO in each direction. Thus we may reduce the number of representers by nearly a factor of ten. Finally, a cheap preliminary calculation of all the remaining representers, using a coarse numerical grid, permits an *array mode analysis* (see §2.5). The analysis shows that a further reduction by a factor of about four is appropriate. In conclusion, about 4000 real representers are needed. They may be fitted to the 1.8×10^6 data values, however.

5.2.8   Constituent dependencies

It might be inferred from the preceding discussion that the representers at different tidal frequencies may be calculated independently. In general, this is not the case. The representer adjoint variables obey

$$\mathbf{S}_k^\dagger \alpha_k = \delta(\mathbf{x} - \boldsymbol{\xi})\hat{\mathbf{e}}_3 \qquad (5.2.33)$$

for $1 \le k \le K$, where $\hat{\mathbf{e}}_3 = (0, 0, 1, 0)^{\mathrm{T}}$, and so may be calculated separately. However the representers obey

$$\mathbf{S}_k \mathbf{r}_k = \sum_{l-1}^{K} \mathbf{C}_{\tau_{kl}} \circ \alpha_l \qquad (5.2.34)$$

for $1 \le k \le K$, and in general the LTE error covariance is not diagonal with respect to k and l. Nevertheless, we may reasonably assume that errors for semidiurnal constituents are independent of those for diurnal constituents. The tidal inverse problem involves immense detail, because so much is known about the structure of the tide-generating force.

5.2.9   Global tidal estimates

Estimating global ocean tides using hydrodynamic models and satellite altimetry is formulated as an inverse problem in Egbert *et al.* (1994). The altimetric data are being inverted in order to find errors in the drag law and bathymetry, especially in the deep ocean. Linear dynamics and linear measurement functionals suffice. The time dependence involves few degrees of freedom and those are highly regular, pure harmonic in fact. The number of cross-over data and hence the number of representers is very large (and still growing, after eight years), yet their number can be reduced by obvious and reasonable subsampling strategies (for example, every third cross-over in each direction), and by a priori array assessment based on economical computation of representers on a coarser numerical grid. The eventual set of decimated and rotated representers may still be fitted closely to the entire data set, however.

Best of all, the challenge of a real, large and important problem led (Egbert, personal communication) to the indirect representer method, outlined in §3.1.3 and applied to real data in Egbert *et al.* (1994; hereafter referenced as EBF). This tidal solution and others have been extensively reviewed (Andersen *et al.*, 1995; Le Provost, 2001; Le Provost *et al.*, 1995; Shum *et al.*, 1997). The solutions were tested with independent tide gauge data. All agreed to within a few centimeters, but the EBF inverse solution ("TPX0.2") did not perform as well as empirical fits to the altimetry (Schrama and Ray, 1994), nor as well as a finite-element forward solution of the Laplace Tidal Equations obtained by a team in Grenoble (Le Provost *et al.*, 1994). The inverse solution was in effect an empirical fit to the altimetry using a few thousand representers, whereas the other empirical fits used around one hundred thousand degrees of freedom. Schrama and Ray (1994) chose the high-resolution Grenoble finite-element solution as the prior, or first-guess for their empirical fit. The prior for the EBF inverse was a finite-difference solution of the Laplace Tidal Equations on a relatively coarse grid. A striking and confidence-enhancing aspect of the inverse solution was its relative smoothness, which it owed to its parsimony or few degrees of freedom. The Grenoble finite-element solution had very fine resolution in shallow seas, where it excelled. The EBF inverse solution was based on representers for cross-overs in deep water only. Driven by the tide-generating force and tidal data at few basin boundaries, the finite-element model is almost a pure mechanical theory and so its success is all the more impressive. More recent implementations (Le Provost, personal communication) have no basin boundaries, that is, the domain is the global ocean and so no tidal data are needed to close the solution. Nevertheless, the tidal solutions are quite accurate. This is a remarkable technical and scientific achievement, surely the most successful theory in geophysics and one of the most successful in all of physics. The finite-element model is limited principally by inaccurate bathymetry and by incomplete parameterizations of drag. It has recently been reformulated as an inverse model, and solved with representers computed by finite-element methods (Lyard, 1999). The latest tidal solutions of various type, now based on eight or more years of altimetry and refined orbit theories, are believed to agree to well within observational errors (e.g., Egbert, 1997). A new independent trial is underway at the time of writing (October 2001). The most recent finite-difference inverse solution (TPX0.4) uses approximately 4×10^4 real valued representers, including many in shallower seas (Egbert and Ray, 2000). A global plot of coamplitude and cophase lines may be found at www.oce.orst.edu/po/research/tide/global.html.

A unique feature of the inverse tidal solutions is the availability of maps of residuals in the equations of motion – the Laplace Tidal Equations. A global plot of the average, per tidal cycle, of the rate of working by the dynamical residuals for the principal lunar semidiurnal constituent M_2 of TPX0.4, is shown in Fig. 5.2.7. Negative values indicate that the tides are losing energy. The largest losses do not occur in regions of the strong boundary currents of the general circulation, such as the Gulf Stream, but instead along the ridges and other steep topography. These errors may be due to the somewhat simplified parameterizations of earth tide and load tide, to unresolved topographic waves or to internal tides. The net loss is a delicate balance involving work done by residuals, by a model bottom drag and by the moon.

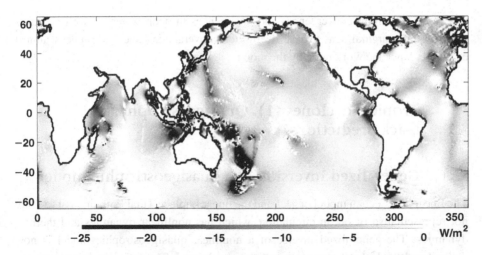

Figure 5.2.7 Per-cycle average rate of working of the M_2 dynamical residuals in TPX0.4 on the M_2 tide, in units of W m^{-2}. Negative values indicate that the M_2 tide is losing energy (after Egbert, 1997).

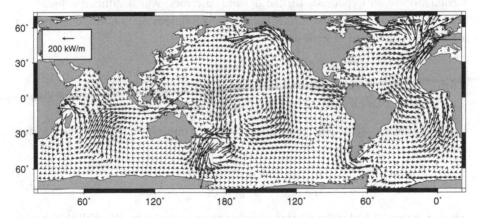

Figure 5.2.8 Flux of total mechanical energy for the linear semidiurnal tidal constituent M_2, based on the inverse model TPX0.4. Note especially the convergence into regions of significant tidal dissipation: for example, the North West Australian shelf, Micronesia/Melanesia and the European shelf (Egbert and Ray, 2001: Estimates of M_2 tidal energy dissipation from TOPEX/POSEIDON altimeter data, *J. Geophys. Res.*, in press. © 2001 American Geophysical Union, reproduced by permission of American Geophysical Union).

The various highly accurate tidal solutions are leading to refined estimates of the dissipation of the energy input to the ocean by the tide-generating force (Lyard and Le Provost, 1997; Le Provost and Lyard, 1997; Egbert and Ray, 2000, 2001). These estimates show that tidal dissipation can provide about 50% of the 2TW of power believed to sustain the meridional overturning circulation, the other 50% being provided by the wind (Wunsch, 1998). A map of energy flux vectors for the tides is shown in Fig. 5.2.8. Some of this power is being produced by the dynamical residuals. Note that there are strong convergences and divergences in deep water, as well as fluxes towards the marginal seas. These deepwater convergences and divergences almost exactly balance the work done by the moon.

5.3   Tropical cyclones (1). Quasigeostrophy; track predictions

5.3.1   Generalized inversion of a quasigeostrophic model

The linear "toy" ocean model of §1.1 and the linear Laplace Tidal Equations of §5.2 are not representative of ocean circulation, which has nonlinear dynamics and thermo-dynamics. The generalized inverse of a nonlinear quasigeostrophic model is now defined and the Euler–Lagrange equations are derived. The latter equations are also nonlinear, so the linear representer algorithm can only be applied iteratively. Two iteration schemes are introduced; a more extensive analysis is provided in §3.3.3. The formulation of an hypothesis for the dynamical errors, which difficult subject was broached §5.2, is considered further here. Finally, implementation of the representer algorithm is discussed in some detail.

5.3.2   Weak vorticity equation, a penalty functional, the Euler–Lagrange equations

Let us formulate weak quasigeostrophic dynamics entirely in terms of the stream-function ψ:

$$\frac{\partial}{\partial t}\nabla^2\psi + \frac{\partial(\psi, \nabla^2\psi + f)}{\partial(x, y)} = \tau, \tag{5.3.1}$$

where τ is the residual in the quasigeostrophic vorticity equation. We shall also specify a weak initial condition for ψ:

$$\psi(\mathbf{x}, 0) = \psi_I(\mathbf{x}) + i(\mathbf{x}), \tag{5.3.2}$$

where i is the initial residual. A weak condition for ψ on the boundary $\mathcal{B}$ of the simply-connected domain $\mathcal{D}$ is

$$\psi(\mathbf{x}, t) = \psi_B(\mathbf{x}, t) + b(\mathbf{x}, t), \tag{5.3.3}$$

where $\mathbf{x}$ lies on $\mathcal{B}$, and b is the boundary streamfunction residual. We shall weakly specify the relative vorticity *all around* $\mathcal{B}$:

$$\nabla^2\psi(\mathbf{x}, t) = \xi_B(\mathbf{x}, t) + z(\mathbf{x}, t), \tag{5.3.4}$$

where $\mathbf{x}$ lies on $\mathcal{B}$ and z is the boundary vorticity residual. See Fig. 5.3.1.

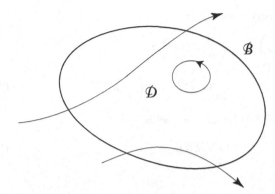

Figure 5.3.1 Planar domain $\mathcal{D}$ with open boundary $\mathcal{B}$. Streamlines cross $\mathcal{B}$ at least twice, or not at all. What can be said about particle paths?

Now (5.3.1) is equivalent to

$$\frac{D\zeta}{Dt} = \frac{\partial \zeta}{\partial t} + \mathbf{u} \cdot \boldsymbol{\nabla} \zeta = \tau, \tag{5.3.5}$$

where $\zeta \equiv \nabla^2 \psi + f$. It follows that if τ is prescribed, then ζ is determined by integrating (5.3.5) along a particle path from an initial position either inside $\mathcal{D}$ or on the boundary $\mathcal{B}$. If any particle path exits $\mathcal{D}$ in the time interval of interest, then (5.3.4) overdetermines ζ. However we shall "adjust" the residuals τ, i, b and z, so that a continuous solution is obtained for ζ, and hence for ψ. More precisely, we shall seek ψ yielding a *smooth* weighted-least-squares best-fit to (5.3.1)–(5.3.4). A suitable penalty functional is (Bennett and Thorburn, 1992):

$$\mathcal{J}[\psi] = \tau \bullet C_\tau^{-1} \bullet \tau + i \circ C_i^{-1} \circ i + b * C_b^{-1} * b + z * C_z^{-1} * z + \mathcal{J}_d, \tag{5.3.6}$$

where $\mathcal{J}_d$ is a penalty for misfits to data within $\mathcal{D}$. Note that

$$\bullet \equiv \int_0^T \!\!\! \iint_{\mathcal{D}} dt\,da, \quad \circ = \iint_{\mathcal{D}} da, \quad * = \int_0^T \!\!\! \iint_{B} dt\,ds, \tag{5.3.7}$$

and all integrations are on a surface of constant pressure p. If $\hat{\psi}$ is a local extremum of $\mathcal{J}$, then

$$\frac{1}{2}\delta\mathcal{J}[\hat{\psi}] = \delta\tau \bullet C_\tau^{-1} \bullet \hat{\tau} + \delta\psi \circ C_i^{-1} \circ \hat{i}$$

$$+ \delta\psi * C_b^{-1} * \hat{b} + \delta\xi * C_z^{-1} * \hat{z} + \frac{1}{2}\delta\mathcal{J}_d = 0. \tag{5.3.8}$$

We shall manipulate the first two terms in detail, leaving the boundaries as an exercise. Define $\hat{\lambda} \equiv C_\tau^{-1} \bullet \hat{\tau}$. Then

$$\delta\tau \bullet \hat{\lambda} = \int_0^T \!\!\! \iint_{\mathcal{D}} dt\,da \left\{ \delta\frac{\partial}{\partial t}\nabla^2\psi + \delta\frac{\partial(\psi, \nabla^2\psi + f)}{\partial(x, y)} \right\} \hat{\lambda}(\mathbf{x}, t). \tag{5.3.9}$$

The first term in (5.3.9) is easily manipulated:

$$\left(\delta\frac{\partial}{\partial t}\nabla^2\psi\right)\hat{\lambda} = \left(\frac{\partial}{\partial t}\nabla^2\delta\psi\right)\hat{\lambda}$$

$$= \frac{\partial}{\partial t}(\hat{\lambda}\nabla^2\delta\psi) - \left(\frac{\partial}{\partial t}\hat{\lambda}\right)\nabla^2\delta\psi. \tag{5.3.10}$$

Integrating the first term in (5.3.10) over time yields

$$[\hat{\lambda}\nabla^2\delta\psi]_{t=0}^T. \tag{5.3.11}$$

Now

$$\hat{\lambda}\nabla^2\delta\psi = \nabla\cdot(\hat{\lambda}\nabla\delta\psi - \delta\psi\nabla\hat{\lambda}) + \delta\psi\nabla^2\hat{\lambda}. \tag{5.3.12}$$

If the area integral of terms proportional to $\delta\psi(\mathbf{x}, T)$ vanishes, for arbitrary values of the latter, then

$$\nabla^2\hat{\lambda} = 0 \quad\text{at } t = T. \tag{5.3.13}$$

Similarly, we infer that

$$-\nabla^2\hat{\lambda} + C_i^{-1}\circ(\hat{\psi} - \psi_I) = 0 \quad\text{at } t = 0. \tag{5.3.14}$$

The second term in (5.3.10) is

$$-\left(\frac{\partial}{\partial t}\hat{\lambda}\right)\nabla^2\delta\psi = -\nabla\cdot\left(\frac{\partial\hat{\lambda}}{\partial t}\nabla\delta\psi - \delta\psi\nabla\frac{\partial\hat{\lambda}}{\partial t}\right) - \delta\psi\nabla^2\frac{\partial\hat{\lambda}}{\partial t}. \tag{5.3.15}$$

The second term in (5.3.15) will be used shortly. Consider now the variation of the Jacobian in (5.3.9):

$$\hat{\lambda}\delta\frac{\partial(\psi, \nabla^2\psi + f)}{\partial(x, y)} = \hat{\lambda}\frac{\partial(\delta\psi, \nabla^2\hat{\psi} + f)}{\partial(x, y)} + \hat{\lambda}\frac{\partial(\hat{\psi}, \delta\nabla^2\psi)}{\partial(x, y)} + O(\hat{\lambda}(\delta\psi)^2) \tag{5.3.16}$$

$$\ldots = -\delta\psi\left\{\frac{\partial(\hat{\lambda}, \nabla^2\hat{\psi} + f)}{\partial(x, y)} + \nabla^2\frac{\partial(\hat{\lambda}, \hat{\psi})}{\partial(x, y)}\right\}$$

$$+ \text{divergence terms.} \tag{5.3.17}$$

So, by requiring the coefficient of $\delta\psi(\mathbf{x}, t)$ to vanish, we recover from (5.3.8), (5.3.15) and (5.3.17) the Euler–Lagrange equation

$$-\frac{\partial}{\partial t}\nabla^2\hat{\lambda} - \nabla^2\frac{\partial(\hat{\psi}, \hat{\lambda})}{\partial(x, y)} = \frac{\partial(\hat{\lambda}, \nabla^2\hat{\psi} + f)}{\partial(x, y)} - \frac{1}{2}\frac{\delta\mathcal{J}_d}{\delta\psi}, \tag{5.3.18}$$

where the last term is a linear combination of measurement kernels. Equation (5.3.18) may be formally rewritten as

$$-\frac{\partial\hat{\lambda}}{\partial t} - \nabla\cdot(\hat{\mathbf{u}}\hat{\lambda}) = \nabla^{-2}\hat{\mu}\cdot\nabla\hat{\zeta} - \frac{1}{2}\nabla^{-2}\left(\frac{\delta\mathcal{J}_d}{\delta\psi}\right), \tag{5.3.19}$$

where $\hat{\mathbf{u}} = \left(-\frac{\partial \hat{\psi}}{\partial y}, \frac{\partial \hat{\psi}}{\partial x}\right)$ and $\hat{\mu} \equiv \left(-\frac{\partial \lambda}{\partial y}, \frac{\partial \lambda}{\partial x}\right)$. Note that $\nabla \cdot \hat{\mathbf{u}} = \nabla \cdot \hat{\mu} = 0$. The form (5.3.19) looks like the "adjoint" of the total vorticity conservation law

$$\frac{\partial \zeta}{\partial t} + \mathbf{u} \cdot \nabla \zeta = \tau, \qquad (5.3.20)$$

but for the emergence of a new term on the rhs of (5.3.19) arising from the variation of the advecting velocity $\mathbf{u}$:

$$\delta \tau = \frac{\partial}{\partial t} \delta \zeta + \mathbf{u} \cdot \nabla \delta \zeta + (\delta \mathbf{u}) \cdot \nabla \zeta. \qquad (5.3.21)$$

No such term arose in our "toy" model $\frac{\partial u}{\partial t} + c \frac{\partial u}{\partial x} = \tau$, since the phase velocity c was fixed. In particular c did not depend upon the state u.

Exercise 5.3.1
Derive the boundary conditions that accompany the variational equation (5.3.18) or (5.3.19). Which boundary condition goes with which equation?   □

5.3.3   Iteration schemes; linear Euler–Lagrange equations

The Euler–Lagrange system (5.3.1) and (5.3.18), with attendant initial and boundary conditions, is nonlinear. The system is coupled through the data term $(\delta \mathcal{J}_d / \delta \psi)$ in (5.3.18), through advection on the lhs of (5.3.18) and through the other term on the rhs of (5.3.18). It is also coupled through the boundary conditions. A simple iteration scheme breaks the coupling completely: calculate a sequence $\{\hat{\psi}^n, \lambda^n\}_{n=1}^{\infty}$ such that

$$\frac{\partial}{\partial t} \nabla^2 \hat{\psi}^n + \frac{\partial(\hat{\psi}^n, \nabla^2 \hat{\psi}^n + f)}{\partial(x, y)} = C_\tau \bullet \lambda^n, \qquad (5.3.22)$$

$$-\frac{\partial}{\partial t} \nabla^2 \lambda^n - \nabla^2 \frac{\partial(\hat{\psi}^{n-1}, \lambda^n)}{\partial(x, y)} = \frac{\partial(\lambda^{n-1}, \nabla^2 \hat{\psi}^{n-1} + f)}{\partial(x, y)} - \frac{1}{2} \frac{\delta \mathcal{J}_d^{n-1}}{\delta \psi}. \qquad (5.3.23)$$

These equations may be solved by integrating (5.3.23) backwards, and (5.3.22) forwards. Note that (5.3.22) is nonlinear in $\hat{\psi}^n$, but the broken coupling eliminates the need for representers! However, such a sequence always seems to diverge.

An alternative iteration scheme is:

$$\frac{\partial}{\partial t} \nabla^2 \hat{\psi}^n + \frac{\partial(\hat{\psi}^{n-1}, \nabla^2 \hat{\psi}^n + f)}{\partial(x, y)} = C_\tau \bullet \lambda^n, \qquad (5.3.24)$$

$$-\frac{\partial}{\partial t} \nabla^2 \lambda^n - \nabla^2 \frac{\partial(\hat{\psi}^{n-1}, \lambda^n)}{\partial(x, y)} = \frac{\partial(\lambda^{n-1}, \nabla^2 \hat{\psi}^{n-1} + f)}{\partial(x, y)} - \frac{1}{2} \frac{\delta \mathcal{J}_d^n}{\delta \psi}. \qquad (5.3.25)$$

This system is *linear*, but coupled: note that the data term in (5.3.25) is evaluated with $\hat{\psi}^n$. It is the Euler–Lagrange system for a linear dynamical model, advected by $\hat{\mathbf{u}}^{n-1}$. There is a first-guess forcing $C_\tau \bullet \lambda_F^n$, where λ_F^n is the response of the lhs of (5.3.25) to the first term on the rhs. The system may be solved using representers. The sequence converges in practice if the Rossby number is moderate. There are some theorems about convergence in doubly-periodic domains: the sequence is bounded and so must have points of accumulation or cluster points, but not necessarily unique limits. There is numerical evidence of the sequence cycling, presumably between cluster points. A third iteration scheme is described in §3.3.3.

5.3.4   What we can learn from formulating a quasigeostrophic inverse problem

A quasigeostrophic inverse model offers some especially clear problems in error estimation. Also, there are special opportunities for reliable estimation of these errors. To the extent that the situation is not representative of Primitive Equation inverse models, one might regard quasigeostrophic inversion as a curiosity but, like the uniquely elegant tidal inverse problem, the quasigeostrophic inverse problem offers valuable experience.

5.3.5   Geopotential and velocity as streamfunction data: errors of interpretation

There are errors of interpretation in certain streamfunction data. Consider geopotentials ϕ and horizontal velocities $\mathbf{u}$ measured by radar-tracking of high altitude balloons, or by sonar-tracking of deeply submerged floats. The quasigeostrophic state variable is the streamfunction field ψ. We must relate ϕ and $\mathbf{u}$ to ψ. The geostrophic approximation is

$$f\hat{\mathbf{k}} \times \mathbf{u} \cong -\nabla\phi, \tag{5.3.26}$$

where the Coriolis parameter is a function of latitude: $f = f(y)$. We have assumed that $\nabla \cdot \mathbf{u} \cong 0$ and that there is a streamfunction for ψ, so (5.3.26) becomes

$$-f\nabla\psi \cong -\nabla\phi. \tag{5.3.27}$$

Ignoring variations in f leads to the "poor man's balance equation"

$$f_0\psi = \phi, \tag{5.3.28}$$

where $f_0 = f(y_0)$ for some latitude y_0. Hence geopotential data may be used as approximations to streamfunction data. Also, velocity data may be used as approximations to streamfunction-gradient data:

$$\nabla\psi = -\hat{\mathbf{k}} \times \mathbf{u}. \tag{5.3.29}$$

Let us begin to estimate the errors in (5.3.28) and (5.3.29). If L is a horizontal length scale and U is a velocity scale, then the local acceleration neglected in (5.3.26) has the scale U^2L^{-1}. The Coriolis acceleration retained in (5.3.26) has the scale f_0U, so the relative errors in (5.3.26) scale as the Rossby number $Ro \equiv \frac{U}{f_0L}$. We shall assume for simplicity that variations in f are smaller than $Ro f_0$. Then (5.3.26) is

$$\hat{\mathbf{k}} \times \mathbf{u} = -f_0^{-1} \nabla \phi + O(U Ro), \qquad (5.3.30)$$

hence

$$\nabla \cdot \mathbf{u} = O\left(Ro \frac{U}{L}\right). \qquad (5.3.31)$$

In general, for any $\mathbf{u}$ there is a streamfunction ψ and a velocity potential χ such that

$$\mathbf{u} = \hat{\mathbf{k}} \times \nabla \psi + \nabla \chi \equiv \mathbf{u}_\psi + \mathbf{u}_\chi, \qquad (5.3.32)$$

thus

$$\xi \equiv \hat{\mathbf{k}} \cdot \nabla \times \mathbf{u} = \nabla^2 \psi, \quad \delta \equiv \nabla \cdot \mathbf{u} = \nabla^2 \chi. \qquad (5.3.33)$$

We conclude from (5.3.31) that χ is $O(Ro\, UL)$ and hence (5.3.29) is accurate to $O(Ro\, U)$, while (5.3.28) is accurate to $O(Ro\, f_0 UL)$. In summary, the "theoretical" relative errors in the data are $O(Ro)$, where $Ro \equiv U/(f_0 L)$. For Gulf Stream meanders in the ocean, $U = 1$ m s^{-1} ($=2$ knots), $L = 10^5$ m and $f_0 = 10^{-4}$ s^{-1}, so $Ro = 0.1$. For middle-level synoptic-scale weather systems in the atmosphere, $U = 30$ m s^{-1} and $L = 10^6$ m, so $Ro = 0.3$.

In the preceding analysis, the estimates of neglected local accelerations were based on the values L and U representative of the synoptic-scale circulation of interest. For consistency, all fields should be low-pass filtered prior to sampling, in order to suppress smaller-scale motions such as internal waves. If, as is often unavoidable, the smoothing is inadequate, then the data will be contaminated with aliased signals. This contamination can be substantial, exceeding for example the estimate $O(RoU)$ for errors in (5.3.29). (I am grateful to Dr Ichiro Fukumori for a discussion of this point. AFB)

5.3.6 Errors in quasigeostrophic dynamics: divergent flow

Estimating the dynamical errors in a quasigeostrophic model is particularly instructive, as we have closed analytical forms for many sources of error. Recall again the momentum balance for the Primitive Equations:

$$\frac{\partial \mathbf{u}}{\partial t} + (\mathbf{u} \cdot \nabla)\mathbf{u} + \omega \frac{\partial}{\partial p}\mathbf{u} + f\hat{\mathbf{k}} \times \mathbf{u} = -\nabla \phi. \qquad (5.3.34)$$

Taking the curl at constant pressure yields

$$\frac{\partial \xi}{\partial t} + (\mathbf{u} \cdot \nabla)\xi + \omega \frac{\partial \xi}{\partial p} + \hat{\mathbf{k}} \cdot \nabla \omega \times \frac{\partial \mathbf{u}}{\partial p} + (f + \xi)\delta + \beta v = 0, \qquad (5.3.35)$$

where $\xi = \hat{\mathbf{k}} \cdot \nabla \times \mathbf{u}, \delta = \nabla \cdot \mathbf{u}$ and $\beta \equiv df/dy$. Splitting $\mathbf{u}$ into a solenoidal part $\mathbf{u}_\psi$ and an irrotational part $\mathbf{u}_\chi$ (see (5.3.31)) leads to a split for (5.3.35):

$$\frac{\partial \xi}{\partial t} + (\mathbf{u}_\psi \cdot \nabla)\xi + \beta v_\psi = -(\mathbf{u}_\chi \cdot \nabla)\xi - (f + \xi)\delta - \beta v_\chi - \omega \frac{\partial \xi}{\partial p} - \hat{\mathbf{k}} \cdot \nabla \, \omega \times \frac{\partial \mathbf{u}}{\partial p} \equiv \tau.$$

(5.3.36)

That is,

$$\frac{\partial}{\partial t} \nabla^2 \psi + \frac{\partial(\psi, \nabla^2 \psi + f)}{\partial(x, y)} = \tau,$$

(5.3.37)

where we have an explicit form for τ in terms of *resolvable fields*. That is, given archives of gridded fields of $\mathbf{u}(\mathbf{x}, p, t)$, we may evaluate τ on the grid, and hence estimate its mean $E\tau$ and covariance C_τ. The most difficult part is calculating ω reliably. We could use the conservation of mass:

$$\frac{\partial \omega}{\partial p} = -\nabla \cdot \mathbf{u},$$

(5.3.38)

subject to $\omega \to 0$ as $p \to 0$, or we could use the conservation of entropy:

$$\omega = \left\{ \frac{\dot{Q}}{T} - \frac{\partial \eta}{\partial t} - \mathbf{u} \cdot \nabla \eta \right\} \left(\frac{\partial \eta}{\partial p} \right)^{-1},$$

(5.3.39)

where $\dot{Q}$ is the heat source per unit mass and T is the absolute temperature. Note that in order to calculate η and T via the equation of state, we need the other thermodynamic state variables such as (p, ρ, q) in the atmosphere, or (p, ρ, S) in the ocean. We may dispense with ρ if T has been measured or is otherwise available on the grid.

There are opportunities to make similar direct estimates of dynamical errors in other "reduced" models, such as balanced models, and the Cane–Zebiak coupled model (Zebiak and Cane, 1987). However, there are additional dynamical errors in all these reduced models, owing to unresolved stresses. The additional errors may exceed the resolvable errors.

5.3.7   Errors in quasigeostrophic dynamics: subgridscale flow, second randomization

We shall consider the unresolved stresses, in the context of the quasigeostrophic vorticity equation

$$\frac{\partial \xi}{\partial t} + \mathbf{u} \cdot \nabla \xi + \beta v = 0,$$

(5.3.40)

where $\xi = \nabla^2 \psi$ and $\mathbf{u} = \hat{\mathbf{k}} \times \nabla \psi$. Note that the subscript "ψ" on $\mathbf{u}$ is now dropped. In practice we can only calculate with (5.3.40) on a grid having some finite resolution in space and time. Yet we know from observations and from instability theory that (5.3.40) possesses solutions that have infinitesimally fine structure of significant amplitude.

We try to separate the coarse and fine structures using the abstraction of an ensemble of flows having a mean (with only the coarse scales), and variability (with only the fine scales). That is, $\xi = \bar{\xi} + \xi'$, where $\bar{\bar{\xi}} = \bar{\xi}$ and $\bar{\xi'} = 0$. In practice we can only approximate the ensemble average (denoted by $\overline{()}$ here) using a space or time average $\widetilde{()}$, but then $\widetilde{(\tilde{\xi})} \neq \tilde{\xi}$. We shall ignore this very important issue here (see Ferziger, 1996 for an excellent discussion) and assume that $\overline{()}$ may be estimated with adequate accuracy. Only the mean field being of interest, it would be desirable to replace the "detailed" vorticity equation (5.3.40) with an equation for $\bar{\xi}, \bar{\mathbf{u}}$ and $\bar{\psi}$. Averaging (5.3.40) yields

$$\frac{\partial}{\partial t}\bar{\xi} + \overline{\mathbf{u} \cdot \nabla \xi} + \beta \bar{v} = 0. \tag{5.3.41}$$

Now for any a and b,

$$\begin{aligned}
\overline{ab} &= \overline{(\bar{a} + a')(\bar{b} + b')} \\
&= \overline{(\bar{a}\bar{b} + \bar{a}b' + a'\bar{b} + a'b')} \\
&= \bar{\bar{a}}\bar{\bar{b}} + \bar{\bar{a}}\bar{b'} + \bar{a'}\bar{\bar{b}} + \overline{a'b'} \tag{!} \\
&= \bar{a}\bar{b} + \bar{a}0 + 0\bar{b} + \overline{a'b'} \\
&= \bar{a}\bar{b} + \overline{a'b'}. \tag{5.3.42}
\end{aligned}$$

Thus

$$\overline{\mathbf{u} \cdot \nabla \xi} = \bar{\mathbf{u}} \cdot \nabla \bar{\xi} + \overline{\mathbf{u}' \cdot \nabla \xi'} \tag{5.3.43}$$

and (5.3.41) becomes

$$\frac{\partial \bar{\xi}}{\partial t} + \bar{\mathbf{u}} \cdot \nabla \bar{\xi} + \beta \bar{v} = \bar{\tau} \equiv -\nabla \cdot (\overline{\mathbf{u}'\xi'}), \tag{5.3.44}$$

where we have used $\nabla \cdot \mathbf{u}' = 0$. So there is another candidate for the residual $\bar{\tau}$ in the *mean* dynamics: the divergence of the mean "eddy-flux" of relative vorticity. Finding a formula for such fluxes in terms of first moments (that is, in terms of $\bar{\xi}, \bar{\mathbf{u}}$ or $\bar{\psi}$) is the *turbulence problem*. It remains unresolved. However, we may use (5.3.44) to constrain the circulation, provided we can put bounds on $\bar{\tau}$. At this point the fast talk begins. Realizing that even the smoothed fields fluctuate considerably, we may regard $\bar{\tau}$ as a random field with a prior mean and variance (prior to assimilating data). Generally we neglect the new mean $E\bar{\tau}$ for $\bar{\tau}$ (or else model it with a diffusion law, for example), and struggle to make scale estimates for the variability in $\bar{\tau}$. For example, if for the eddies $|\bar{\mathbf{u}}| \sim U$ and $|\mathbf{x}| \sim l$, we might be tempted to assume $\bar{\tau} \sim U^2 l^{-2}$. This is usually excessive; the length scale L of the (smoothed) eddy-flux $\overline{\mathbf{u}'\xi'}$ is much greater than the length scale l of the eddies themselves. That is, $\bar{\tau} \sim cU^2 L^{-1}l^{-1}$, where $c \ll 1$ is the magnitude of the correlation coefficient between $\mathbf{u}'$ and ξ'. The decorrelation length scale $\mathcal{D}$ for $\bar{\tau}$ presumably lies in the interval $l < \mathcal{D} < L$, while the decorrelation time $\mathcal{T}$ lies in the range $(l/U) < \mathcal{T} < (L/U)$. In the jet stream or ocean boundary currents, on the other hand, $l \sim L$. For the weakly homogeneous case ($l \ll L$), however, we might hypothesize that

$$E(\overline{\tau}(\mathbf{x}, t)\overline{\tau}(y, s)) = C_{\overline{\tau}}(\mathbf{x}, t, y, s) = \frac{c^2 U^4}{L^2 l^2} \exp\left\{-\frac{|\mathbf{x} - \mathbf{y}|^2}{\mathcal{D}^2} - \frac{|t - s|^2}{T^2}\right\}. \quad (5.3.45)$$

One might reasonably feel uncomfortable at this point, attempting to constrain a circulation estimate with such a speculative hypothesis. Indeed, the "second randomization" of $\overline{\tau}$ is a naïve abstraction of the hoped-for scale separations in the fluctuations in $\overline{\tau}$. One should recall that the conventional forward model is merely a circulation estimate based on the hypothesis that $\overline{\tau} \equiv 0$. This is the one hypothesis that we know immediately to be wrong. We could abandon the concept of an ensemble of mean vorticity fluxes, and just manipulate $\overline{\tau}$ as a control that guides the state towards the data. The Euler–Lagrange equations of the calculus of variations enable the manipulations, once a penalty functional has been prescribed. The difficulty lies in the choice of weights. Probabilistic choices (inverses of covariances) are conceptually shaky. Yet the prospect of an ocean model as a testable hypothesis is so appealing.

It was established in §2.2 that generalized inversion is equivalent to optimal interpolation in space and time. The former requires the dynamical error covariance C_{τ}; the latter requires the circulation or state covariance such as C_{ψ}. Which is the easier to specify a priori? We anticipate that ψ is nonstationary, anisotropic and significantly inhomogeneous. The components of multivariate circulation fields will be jointly covarying. On the other hand, it is plausible that the dynamical residuals in *unreduced* models are the result of small-scale processes that are locally stationary, isotropic and univariate. Then the generalized inverse constructs highly structured state covariances guided by the model dynamics, and by the morphology of the domain: the orography, or the bathymetry and coastline.

5.3.8   Implementation; flow charts

The linear representer method is complicated. Its iterative application to a nonlinear quasigeostrophic model makes it even more complicated. Some general suggestions on implementation are in order.

(i) Start with a simple, linear problem first, such as the one described in §1.1–§1.3. The computing exercises at the end of this book provide numerical details. FORTRAN code is available from an anonymous ftp site:
`ftp.oce.orst.edu, cd/dist/bennett/class`.

(ii) A flow chart for the "quasigeostrophic inverse" is given in Figs. 5.3.2 and 5.3.3. The latter figure shows in detail the hatched section in the former. These computations are manageable using a workstation. Your code should consist of a main program that calls many subroutines. These should include a single "backward integration" and a single "forward integration". Preconditioned conjugate gradient solvers are widely available in subroutine libraries.

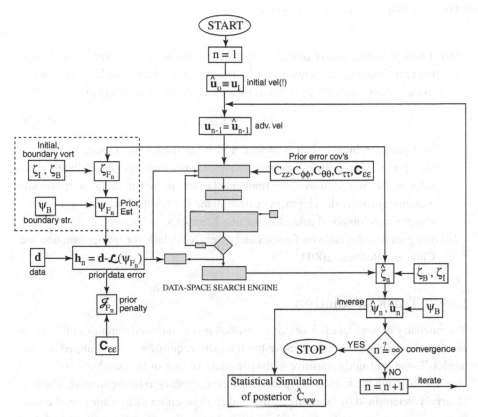

Figure 5.3.2 Generalized inversion of a regional quasigeostrophic model; indirect representer algorithm.

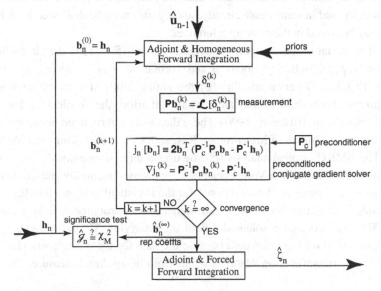

Figure 5.3.3 Data-space search engine.

(iii) The representer matrix should be tested for symmetry. The optimal values $\hat{\beta}$ for the representer coefficients, found by the gradient solver, should be compared to the values available a posteriori by measuring the inverse $\hat{u}(x, t)$:

$$\hat{\beta} = -\tilde{w}(\mathcal{L}[\hat{u}] - \mathbf{d}). \tag{5.3.46}$$

(iv) Very large problems, such as those described in the following section, require very powerful computers having massive memory and disk. It is difficult to offer further suggestions about implementations as each manufacturer provides a unique software development environment. The results presented in this chapter were obtained using Connection Machines.

(v) For pseudocode, code on ftp sites and extensive details for implementation see Chua and Bennett (2001).

5.3.9   Track prediction

The intensity of tropical cyclones[1] is controlled by the thermodynamics of the atmosphere and ocean together. Predicting the intensity requires a fully stratified coupled model. These are highly sensitive to the parameterization of heat exchange (Emanuel, 1999). Predicting the track of a typhoon, however, appears to be far simpler. The track is largely determined by "steering" winds, taken to be either an average over the fields in mid-troposphere, or else the fields at 500 mb. In either case, the evolution of these winds can be represented for a short time (say, one-third of a synoptic time scale, or about a day) by single-level, quasigeostrophic dynamics. These simplified dynamics are nonetheless nonlinear, and so provide a relatively simple yet real and motivated first test for time-dependent variational assimilation in a nonlinear model. The formulation has been discussed in some depth already, so only data need be discussed here. Further details may be found in the ensuing references.

The state variable for the quasigeostrophic model (5.1.38) is the streamfunction $\psi = \psi(x, y, p, t)$; isobaric velocity $\mathbf{u}$ and vorticity ξ may be derived from it: see (5.1.33), (5.1.34). Observations through the entire depth of atmosphere were collected during a typhoon season by an international effort, the Tropical Cyclone '90 or TCM-90 experiment (Elsberry, 1990). These data were interpolated onto regular grids by the Australian Bureau of Meteorology Research Centre (Davidson and McAvaney, 1981). The BMRC tropical analysis scheme uses a three-dimensional univariate statistical interpolation method. Vortex centers were inserted manually and synthetic profiles were used to generate "observations" for the statistical analyses (Holland, 1980). The gridded velocities were then partitioned into a rotational field $\mathbf{u}_\chi$ satisfying $\hat{\mathbf{k}} \cdot \nabla \times \mathbf{u}_\chi = 0$, and a solenoidal field $\mathbf{u}_\psi$ satisfying $\nabla \cdot \mathbf{u}_\psi = 0$. The gridded streamfunction ψ for the latter field became the data for the quasigeostrophic assimilation. These streamfunction "data" were far from being direct measurements of the

---

[1] Or typhoons, as they are known in the Pacific ("hurricanes" in the Atlantic).

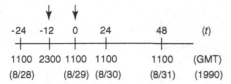

Figure 5.3.4 Time line for generalized
inversion of the tropical cyclone model.
Initial data are available at $t = -24$ in the
form of TCM-90 analyses. Boundary data
are available for $-24 \leq t \leq 0$, also from
TCM-90 analyses. For $0 \leq t \leq 48$, the
boundary data are provided by a global
forecast of relatively coarse resolution. The
inverse takes advantage of additional
TCM-90 data, at $t = -12$ hours and at
$t = 0$ hours (as indicated by arrows). The
dates refer to TC "Abe".

atmosphere. They were further modified by a projection onto the leading ten empirical
orthogonal functions (EOFs), which captured 94% of the variance. The time line for
these derived data, and for the assimilation-forecast episode is shown in Fig. 5.3.4. The
gridded streamfunction data at $t = -24$ hours constitute the prior initial condition, ten
EOF amplitudes were admitted at $t = -12$ and also at $t = 0$, and the smoothing or
inversion interval was $-24 \leq t \leq 0$. The gridded inverse streamfunction at time zero,
that is, $\hat{\psi}(\mathbf{x}, 0)$, became the initial condition for a forward integration or "forecast"
out to $t = +48$. The forecast was honest: boundary data for $0 \leq t \leq 48$ were obtained
from a global forecast model also starting at $t = 0$, rather than from archived analyses.
Ten cases were considered; some involved the same typhoon in different stages of its
life. Detailed results may be found in Bennett *et al.* (1993).

From a scientific perspective the most interesting result is that the values of the
reduced penalty functional $\hat{\mathcal{J}}$ were broadly in the range 20 ± 6, as expected for χ_{20}^2.
Thus the hypothesized error covariances were consistent with the data. A diagnosis
showed that the dynamical residuals and boundary vorticity residuals were negligible,
so it was the hypothesized error covariances for the initial conditions and data that were
consistent with the data.

From a control-theoretic perspective the most interesting result is that there were
sufficiently many degrees of freedom in the initial residuals at $t = -24$ to "aim" the
model at the few data, without additional "guidance" from dynamical residuals for
$-24 \leq t \leq 0$.

From a forecasting perspective the most interesting result was the skill enhance-
ment relative to other track prediction methods: see Fig. 5.3.5. Forty-eight-hour track
predictions based on variational assimilation over the preceding 24 hours were always
superior to those based on either a carefully tuned "nudging" scheme and/or a purely
statistical scheme (Bennett, Hagelberg and Leslie, 1992).

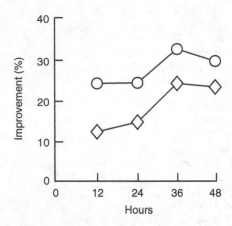

Figure 5.3.5 Tropical cyclone track predictions. Percentage improvement in (reduction of) mean forecast error to 48 hours ahead, relative to climatology plus persistence (CLIPER). Diamonds: standard initialization; circles: initialization by generalized inversion. By convention the 7% difference at 48 hours, for example, is scaled with respect to the 70% residual yielding a "10% reduction".

5.4   Tropical cyclones (2). Primitive Equations, intensity prediction, array assessment

The evolution of a tropical cyclone is a thermodynamic process. Quasigeostrophic dynamics assume that the stratification remains close to the mean, and such is not the case in a tropical cyclone. The Primitive Equations (see §5.1) include laws for:

  (i) conservation of mass in a fully compressible gas, (5.1.9);
 (ii) conservation of relative humidity q, (5.1.24);
(iii) conservation of entropy η, (5.1.24); and
 (iv) an equation of state, (5.1.26), relating density ρ to η, to q and to the pressure p.

The state variable η may be replaced with the temperature T defined by the combined first and second laws of thermodynamics: $\left(\frac{\partial \eta}{\partial T}\right)_p = C_p/T$, $\left(\frac{\partial \eta}{\partial T}\right)_v = C_v/T$, $\left(\frac{\partial \eta}{\partial v}\right)_T = p/T$, where $v = \rho^{-1}$ is the specific volume, while C_p and C_v are respectively the specific heats at constant pressure and volume. For a dry, calorifically perfect (C_p, C_v constant) ideal ($\eta = C_v \ln(p\rho^{-C_p/C_v})$) gas,[2] $T = T_0\left(\frac{\rho}{\rho_0}\right)^{\gamma+1} e^{\eta/C_v}$. Empirical corrections may be made for moisture: see, e.g., Wallace and Hobbs (1977).

---

[2] Boltzmann's equation leads directly to this definition of an ideal gas in terms of its entropy dependence, rather than in terms of the gas law $p = R\rho T$. The latter merely defines temperature. See, e.g., Chapman and Cowling (1970).

The full Primitive Equations may be found in Haltiner and Williams (1980, p. 17) or in the appendix to Bennett, Chua and Leslie (1996, hereafter BCL1; the associated Euler–Lagrange equations are also here in Appendix B). The vertical coordinate is not simply the pressure p as in §5.1, but Phillip's sigma-coordinate: $\sigma = p/p_*$ where p_* is the pressure at the earth's surface. The lower boundary for the atmosphere is conveniently located at $\sigma = 1$.

A quadratic penalty functional for reconciling dynamics, initial conditions and data is also given in BCL1, along with

  (i) the nonlinear Euler–Lagrange equations;
 (ii) the linearized Primitive Equations and Euler–Lagrange equations;
(iii) the representer equations and
(iv) the adjoint representer equations.

The linearized equations (ii)–(iv) enable an iterative solution of the nonlinear equations (i); each linear iterate may itself be solved by the indirect, iterative representer method described in §3.1. The "inner" or "data space" search was preconditioned in BCL1 using all representers calculated on a relatively coarse $128 \times 64 \times 9$ global grid with two-minute time steps, see Bennett, Chua and Leslie (1997, hereafter BCL2). The smallness of the time steps is due to the polar convergence of the meridians. The inverse was calculated on a relatively fine $256 \times 128 \times 9$ global grid with one-minute time steps. There were 4.4×10^8 grid points in a twenty-four-hour smoothing interval, for about 2.6×10^9 gridded values of u, v, $\dot{\sigma}$, T, q, $\ln p_*$, etc.

The coarse-grid preconditioner was only moderately effective owing to errors of interpolation from the coarse grid to the data sites. The latter were reprocessed cloud track wind observations (RCTWO) inferred from consecutive satellite images of middle and upper-level clouds (Velden *et al.*, 1992). Some of these observations are shown in Fig. 5.4.1. The observation period included tropical cyclone "Ed" near (113°E, 15°N) and Supertyphoon[3] "Flo" near (130°E, 23°N). The RCTWO were available at $t = -24$, -18, -12 and 0 hours, and at 850 hPa, 300 hPa and 200 hPa for a total of $M = 2436$ vector components. The measurement errors for each component were assumed to be $3 \, \text{m s}^{-1}$, $4 \, \text{m s}^{-1}$ and $4 \, \text{m s}^{-1}$ at the respective levels, uncorrelated from the other component of the same vector and from all other vectors elsewhere and at different times. The single inversion reported in BCL1 reduced the penalty functional from a prior value of 6432 to a posterior value of 4066. It may be concluded that the forward model and initial conditions (an ECMWF analysis) were very good, that the RCTWO only had moderate impact, and that the prior root mean square error should have been 30% larger. Given the difficulty in estimating the dynamical errors, such a conclusion is incontestable. Assimilation of the RCTWO did however have a useful impact on subsequent forecasts of meridional wind fields near "Flo": see BCL1. Of greater interest here are the representers, for the Primitive Equation dynamics linearized

---

[3] According to the Japanese Meteorological Agency.

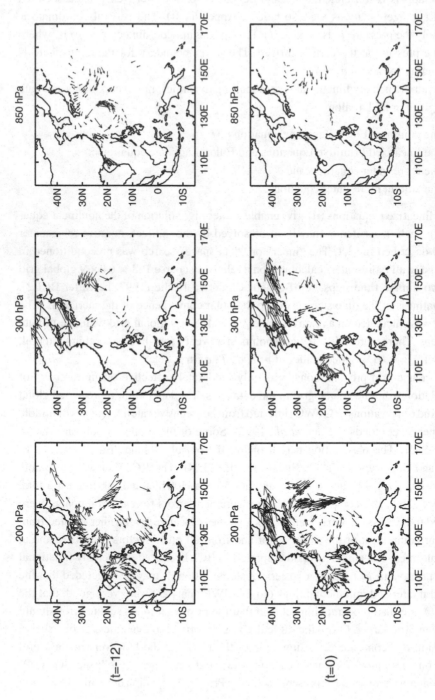

Figure 5.4.1 Reprocessed cloud track wind observation vectors at 200 hPa, 300 hPa and 850 hPa, at $t = -12$ or 0000UTC on 16/IX/1990 (upper panels), and at $t = 0$ or 0000UTC on 16/XI/1990 (lower panels). RCTWO data are also available at $t = -24$, and at $t = -18$, for a total of 2436 scalar data (after Bennett, Chua and Leslie, 1996).

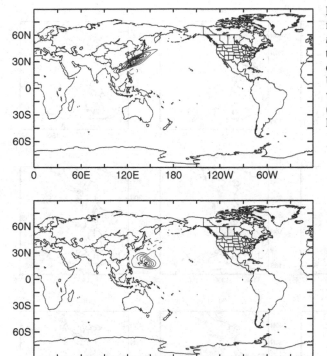

Figure 5.4.2 Zonal velocity fields at ($p = 200$ hPa, $t = 0$) for representers of two RCTWO zonal components also at (200,0). Units are m^2 s^{-2}, as for velocity autocovariances (after Bennett, Chua and Leslie, 1997).

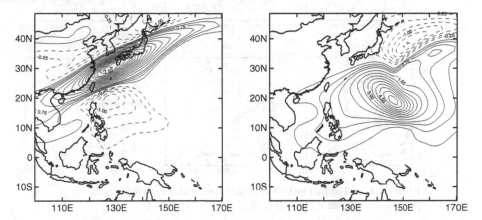

Figure 5.4.3 As for Fig. 5.4.2; close-ups over the S. China Sea (after Bennett, Chua and Leslie, 1997).

about the fifth and final "outer" iterate of the inverse estimate. Upper level zonal winds for two representers are shown in Fig. 5.4.2, and in close-up in Fig. 5.4.3. Their striking anisotropy is a consequence of shearing by supertyphoon winds. The eigenvalues of the $M \times M$ representer matrix $\mathbf{R}$ and its stabilized form $\mathbf{P} = \mathbf{R} + \mathbf{C}_\epsilon$, where $\mathbf{C}_\epsilon$ is

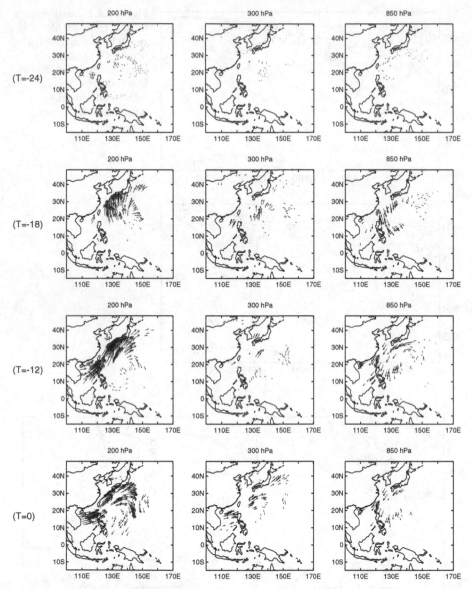

Figure 5.4.4a The first orthonormalized eigenvector $\mathbf{z}_1$ of the symmetric positive-definite matrix $\mathbf{P} = \mathbf{R} + \mathbf{C}_\epsilon$.

the data error covariance matrix, may be seen in BCL1. Recall that $\mathbf{R}$ was calculated on the relatively coarse grid as a preconditioner for indirect inversion on the fine grid. Given the assumed levels of data error, there are only about 200 effective degrees of freedom in the observations. Thus only about 200 iterations would be needed in order to solve (3.1.6). The coarse grid precondition reduced the number to below 15.

   The first and fourth leading normalized eigenvectors of $\mathbf{P}$ are shown in Fig. 5.4.4. They are associated with the first and fourth largest eigenvalue of $\mathbf{P}$: see §2.5 and

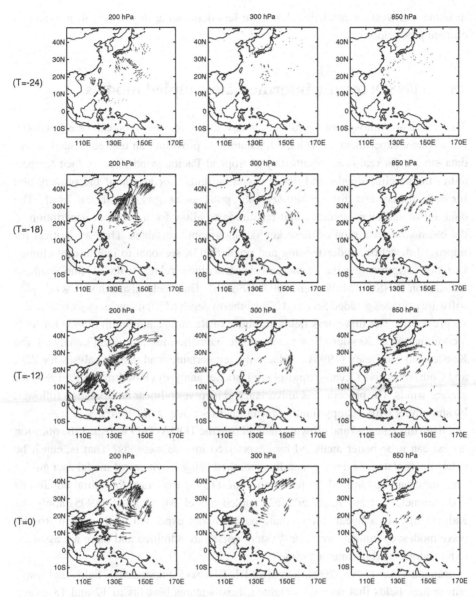

Figure 5.4.4b The fourth eigenvector z_4.

especially (2.5.4). These are the first and fourth most stably observed wind patterns. There is cross-correlation between wind components u and v; there is autocorrelation in time and there is autocorrelation in height. The amplitudes of the winds are asymmetric with respect to the centers of the tropical cyclones, evidently as a consequence of strongly asymmetrical advection. The two eigenvectors display markedly different flow topologies. Variational methods are capable of extracting non-intuitive covariance structure from dynamics, even if the use of such methods for actual assimilation or analysis cannot be afforded in real time. The real-time imperative is most demanding in numerical weather prediction, but is far less demanding in seasonal-to-interannual climate prediction.

5.5   ENSO: testing intermediate coupled models

The Tropical Atmosphere–Ocean array developed for the Tropical Ocean–Global Atmosphere experiment, or TOGA-TAO array, is providing an unprecedented *in situ* data stream for real-time monitoring of tropical Pacific winds, sea surface temperature, thermocline depths and upper ocean currents. For a tour of the project, and for data display and distribution, see www.pmel.noaa.gov/tao/home.html. The data are of sufficient accuracy and resolution to allow for a coherent description of the basin scale evolution of these key oceanographic variables. They are critical for improved detection, understanding and prediction of seasonal to interannual climate variations originating in the Tropics, most notably those related to the El Niño Southern Oscillation (ENSO) (McPhaden, 1993, 1999a,b). The freely-distributed TAO display software provides gridded SST and 20° isotherm depth (Z20) using an objective analysis procedure. The first-guess fields are those of Reynolds and Smith (1995) for SST; a combination of Kessler (1990) expendable bathythermograph (XBT) analyses and Kessler and McCreary (1993) conductivity, temperature, and depth analyses for Z20, and Comprehensive Ocean–Atmosphere Data Set analyses (Woodruff *et al.*, 1987) for surface winds. The procedure is univariate and involves bilinear interpolation followed by smoothing with a gappy running mean filter (Soreide *et al.*, 1996).

Given this splendid and growing data set (see the TOGA-TAO website), the question arises: can it be better analyzed by generalized inverse methods? That is, can it be better interpolated, or more generally smoothed using a dynamical model as a guide? The question is addressed by Kleeman *et al.* (1995) who vary the initial conditions and parameters of an "intermediate" coupled model. Miller *et al.* (1995) apply the Kalman filter to a linear intermediate ocean model expanded in its natural Rossby wave modes; dynamical errors or "system noise" are admitted and these are assumed to be uncorrelated in time or "white".

Bennett *et al.* (1998, 2000, hereafter BI, BII) seek upper ocean fields and lower atmosphere fields that provide weighted, least-squares best-fits to 12 and 18 month segments of monthly mean TAO data, and to a nonlinear intermediate coupled model after that of Zebiak and Cane (1987). The model structure is indicated schematically in Fig. 5.5.1; the equations of motion may be found in the references. The dynamical variables are anomalies of current, wind temperature and layer thickness, relative to their respective annual cycles. The oceanic and atmospheric dynamics are linear, save for the presence of anomalous advection of anomalous heat in the oceanic upper layer, for the quadratic dependence of anomalous surface stress upon anomalous wind, and for the parameterization of turbulent vertical mixing in the ocean in terms of a

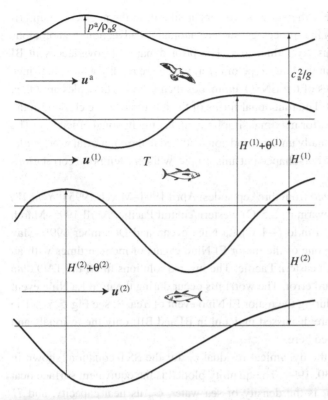

Figure 5.5.1 A reduced gravity, two-and-one-half layer ocean model coupled to a reduced gravity, one-and-one-half layer atmospheric model (after Bennett *et al.*, 1998).

piecewise differentiable switching function. The ocean model domain is a rectangle on the equatorial beta plane: (123.7°E, 84.5°W) × (29°S, 29°N). The atmospheric model domain is an entire equatorial zone (29°S, 29°N). The inclusion of local accelerations in all the momentum equations permits the satisfaction of rigid meridional boundary conditions in the ocean, and the satisfaction of rigid zonal boundary conditions in both the ocean and atmosphere. The inclusion of pseudoviscous stresses permits the satisfaction of no slip and free slip at meridional and zonal boundaries respectively.

The generalized inverse of this intermediate coupled model and the TAO data is, again, the weighted least-squares best fit to the dynamics, the initial conditions and the data. The weights are, as usual, the operator-inverses of the covariances of the dynamical, initial and observational errors. The three error types are assumed mutually uncorrelated. The root mean square data errors are: 0.3° for Sea Surface Temperature (SST), 3 m for the 20° isotherm depth (Z20) and 0.5 m s⁻¹ for each wind component (u^a, v^a). The initial errors are assigned the covariance parameters of the ENSO anomalies themselves (see Kessler *et al.*, 1996), and are assumed mutually uncorrelated. Most difficult of all is the prescription of dynamical error covariances. There will inevitably be errors in the parameterizations of turbulent mixing and exchange processes. In the case of intermediate models, there are also errors arising from the neglect of numerically resolvable pseudolaminar processes, such as anomalous advection of anomalous momentum and anomalous layer thickness. The prior dynamical covariances in BI and BII are based solely on the latter type of error, which are readily assessable since these would have the scales of the ENSO anomalies themselves. The scales are taken from Kessler *et al.* (1996). The functional forms of the covariances are chosen, in the absence of real knowledge, for maximal simplicity and computational efficiency. The variances are stationary, zonally uniform and concentrated in the equatorial waveguide. The spatial correlations are bell-shaped but anisotropic, while the temporal correlations are Markovian (see §3.1.6).

The TAO data are selected from three episodes: April 1994–March 1995 ("Year 1") covering an anomalously warm (+2.5°C) western/central Pacific; April 1995–March 1996 ("Year 2") covering a mild (−1°C) La Niña event, and December 1996–May 1998 ("Year 3"), covering one of the major El Niño events of modern times with an anomalously warm (+5°C) eastern Pacific. The inverse solutions fit all the TAO data to within about one standard error. The worst fits occur during the mild La Niña event of Year 2; the best occur during the major El Niño event of Year 3: see Fig. 5.5.2. The inverse circulation fields are discussed in detail in BI and BII; only the residuals and diagnostics will be reviewed here.

Consider for example the dynamical residual r_T for the SST equation, shown in Fig. 5.5.3 for September 30, 1994. The quantity plotted is the equivalent surface heat flux $\rho_1 C_p H r_T$, where ρ_1 is the density of sea water, C_p its heat capacity, and H the thickness of the ocean surface layer. The contour interval is 20 W m^{-2}. The prior estimate of 50 W m^{-2} is very significantly exceeded over large regions, mostly on the equator. The zonal scale of 30° is that of the corresponding covariance. This field of residuals is one day's distribution of heat sources and sinks that must be admitted in the model if the local rate of change of SST is to be consistent with the TAO data. There are two candidates for r_T: the unresolved advective heat fluxes (both horizontal and vertical), and the missing heat exchange between the model ocean and the model atmosphere. The atmospheric component of the coupled model exchanges heat with the oceanic component at the rate $\dot{Q}_S$, but not vice versa. Radiative feedback from clouds is thereby excluded. The atmospheric budget for geopotential anomaly ϕ is of the form

$$\frac{\partial \phi}{\partial t} \cdots = -\dot{Q}_S = -KT, \qquad (5.5.1)$$

where T is the SST anomaly and K is a positive constant. Thus a positive SST anomaly (and therefore atmospheric heating) leads to a decrease in geopotential anomaly. The region of significant and positive r_T on Sept. 30, 1994 coincides with a positive anomaly T (see BI, Fig. 5). Hence both the model ocean *and* atmosphere gain heat locally on that day. It must be concluded that r_T represents mostly an unresolved convergence

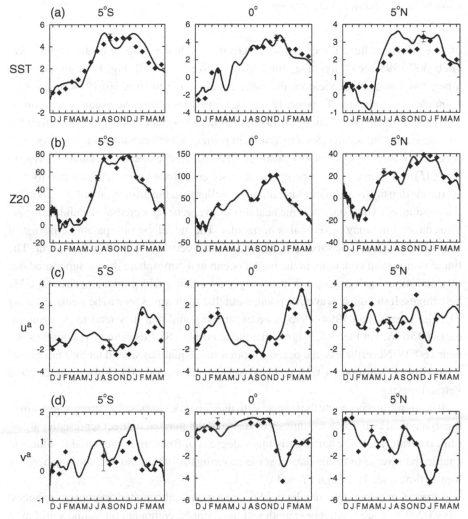

Figure 5.5.2 Time series of the inverse estimate of the anomalous state at three TAO moorings (5°S, 0°, 5°N) along 95°W. The centered symbols are the 30-day average TAO data. All data of the same type are assigned the same standard error so only one bar is shown per panel, but note that the amplitude scale and bar length vary from panel to panel. Results here are for "Year 3" (December 1996–May 1998): (a) SST, ± 0.3 K; (b) Z20, ± 3 m; (c) u^a, ± 0.5 m s^{-1}; (d) v^a, ± 0.5 m s^{-1} (after Bennett *et al.*, 2000).

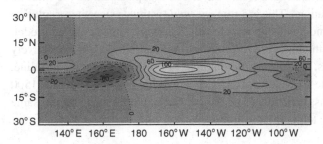

Figure 5.5.3 Scaled residual for the temperature equation for day 180, Year 1 (September 30, 1994). Contour interval: 20 W m^{-2} (after Bennett *et al.*, 2000).

of oceanic heat flux, rather than a neglected exchange with the atmosphere. On "Feb. 30" 1995, the region of significantly negative r_T (see BI, Fig. 12) coincides with a negative anomaly T, yielding the same conclusion. On Nov. 30, 1994 negative r_T coincides with positive T, possibly representing a loss from the ocean to the atmosphere rather than an oceanic heat flux divergence. No clear evidence of loss from the atmosphere to the ocean is seen in Year 1. In principle, both candidates for r_T should be accepted. However, scale analysis shows that an oceanic temperature source of strength $(c_a^2/gH)\rho^a K T(\rho_1 C_p)^{-1}$, c_a being the atmospheric phase speed, g being gravity and ρ^a the air density, is an order of magnitude smaller than the prior standard deviation for the residual r_T. Thus, only oceanic heat flux convergence is a credible candidate for r_T. The convergence may be vertical or horizontal. The model's simple parameterization of heat flux using a simple mixing function is almost certainly significantly in error. The linear momentum equations in the model ocean and atmosphere do not support eddies or instabilities such as tropical instability waves that could produce horizontal eddy heat fluxes. It should, however, be pointed out that such waves tend to be weakest during El Niño events (and 1994–1995 is no exception), and that they tend to be strongest east of 150°W. Yet Fig. 5.5.3 shows that the maximum SST dynamical residuals r_T are near 160°W. Nevertheless, the oceanic momentum equations should include horizontal advection, in addition to well-resolved vertical advection and better-parameterized vertical mixing.

It is simple to recompute the inverse with the dynamics imposed as strong constraints: the dynamical error variances are set to zero and the iterated indirect representer algorithm is rerun. There are sufficiently many degrees of freedom in the initial residuals to enable the inverse to fit the data at some moorings for three months, but nowhere for longer times: see Fig. 5.5.4 (Year 1).

Monte Carlo methods may be used to approximate the posterior error covariances: see §3.2. These are relatively smooth and need not be computed on as fine a grid as is used for the inverse itself. A small number of samples should be adequate for such low moments of error, if not for Monte Carlo approximation of the inverse itself. Recall that the representers are themselves covariances (see §2.2.3), and so may be approximated by Monte Carlo methods. Comparisons with representers and inverses calculated with the Euler–Lagrange equations demonstrate the accuracy of sampling methods. Shown in Fig. 5.5.5 are four calculations of SST for Nov. 1994. Daily values are calculated as described below, and then averaged for 30 days. The first panel shows the solution of the Euler–Lagrange equations. This is a true *ensemble* estimate since it is a solution to what are, in effect, the moment equations for the randomly forced coupled model. The second, third and fourth panels are Monte Carlo estimates based on respectively 100, 500 and 1500 samples. It is disturbing that the $+2°$ warm pool on the Dateline, characterizing the moderate El Niño of Year 1, is only clearly expressed with 1500 samples. These calculations, variational and Monte Carlo, are all made on the same spatial grid and at the same temporal resolution.

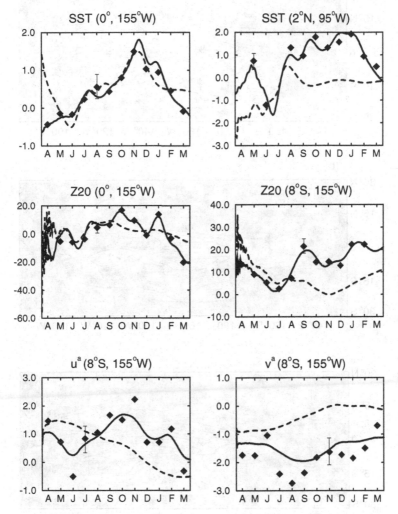

Figure 5.5.4 TAO 30-day averaged data (centered symbols) and time series of SST, Z20, u^a and v^a at selected moorings. Solid lines: weak constraint inversion; dashed lines: strong constraint inversion (Year 1, after Bennett *et al.*, 1998).

Monte Carlo approximation of the error covariances indicates a more relaxed state of affairs. Shown in Fig. 5.5.6 are the prior and posterior variances of initial SST errors as functions of longitude and latitude. The great difference between the prior and posterior (or "explained") variances, with 140 samples, greatly exceeds the sampling error in the prior variance. The small posterior variance implies that the initial SST estimate is reliable. Similar implications hold for the inverse estimates of SST throughout Year 1: see BI, Figs. 17 and 18 for prior and posterior variances for that variable and other coupled model variables. There is, however, a caveat. All these priors and posteriors

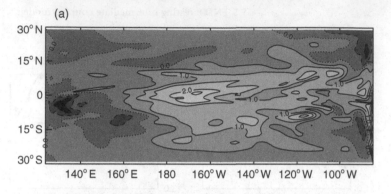

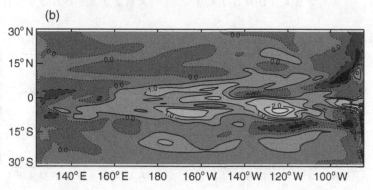

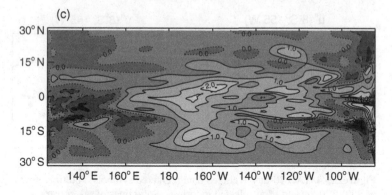

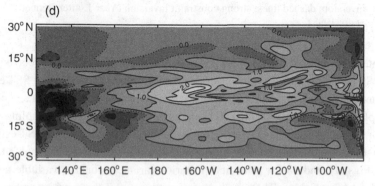

Figure 5.5.5 Inverse estimate of daily SST averaged for month 8, Year 1
(Nov. 1994); (a) as a solution of the Euler–Lagrange equations, (b) as a
statistical simulation with 100 samples, (c) 500 samples, (d) 1500 samples.

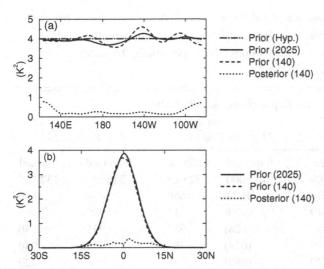

Figure 5.5.6 Statistical simulations of (a) equatorial and (b) meridional profiles of prior and posterior error variances for initial SST. The level broken line in (a) is the hypothesized initial equatorial error variance of $4K^2$. In (b), the hypothesis is indistinguishable from the solid line. The numbers in parentheses indicate the number of samples (after Bennett *et al.*, 1998).

are based on the null hypothesis for the prior errors in the initial conditions, data and dynamics. Thus the posteriors derived from the hypothesized priors cannot be trusted until the null hypothesis has survived significance tests.

The prior and posterior values of the penalty functional are test statistics for the null hypothesis: see §2.3.3. Their values for the three El Niño–La Niña episodes are shown in Table 5.5.1, which is taken from BII. They are calculated using, where needed, a Monte Carlo estimate of the full representer matrix **R**. Note that the prior $\mathcal{J}_F$ and posterior $\hat{\mathcal{J}}$ need only the prior data misfit vector **h**, the specified measurement error covariance matrix $\mathbf{C}_\epsilon$, and the representer coefficient vector $\hat{\beta}$ which may be obtained without explicit construction of **R**: see §3.1.4. With the exception of $\hat{\mathcal{J}}$, the expectations and variances of these statistics all do depend explicitly upon **R**. That the actual values of $\mathcal{J}_F$ in all three "Years" are significantly less than their expected values, suggests that the forward model is far more accurate than hypothesized. Inversion would seem unnecessary. Yet, the actual values of $\hat{\mathcal{J}}$ for the three years exceed their expected values by 15, 16 and 49 standard deviations, respectively. On the other hand, rescaling the standard deviations of the errors in the null hypothesis by 1.40, 1.44 and 2.08, respectively, would yield values of $\hat{\mathcal{J}}$ equal in each case to the expected value M given by the number of data. Such rescalings could hardly be contested, in light of the uncertainties involved in developing the null hypothesis. A fourth year of data, for another El Niño event, is needed in order to obtain at least one independent test of the last upward rescaling. It would serve little purpose, as the dynamical residuals already dominate the term balances: see Fig. 5.5.7. A rescaling of the priors might well yield a statistically self-consistent analysis of TAO data using an intermediate coupled model, but the model constraint would be so "slack" that it would provide no dynamical insight into ENSO. Fully stratified models are needed, with fine vertical resolution and good estimates of moments of errors in the turbulence parameterizations.

The calculations described above involve about 4×10^7 control variables or residuals; there are about 2500 monthly-mean data in the 12-month episodes (Years 1 and 2)

Table 5.5.1 (a) Expected and actual values of components of the reduced penalty functional for the intermediate coupled model (those values in parentheses are numbers of data rather than expected values of data penalties); (b) standard deviations (after Bennett *et al.*, 2000).

(a) Expected and actual values

	Year 1 $M = 2624, \sqrt{2M} = 72$		Year 2 $M = 2644, \sqrt{2M} = 73$		Year 3 $M = 4008, \sqrt{2M} = 89$	
	Expected	Actual	Expected	Actual	Expected	Actual
$\mathcal{J}_F$	16 0000	35 708	15 6000	32 028	246 132	132 140
$\mathcal{J}_{mod}$	1015	1614	1022	1650	1458	3680
$\mathcal{J}_{SST}$	(689)	419	(700)	717	(1088)	995
$\mathcal{J}_{u^a}$	(624)	453	(628)	383	(931)	1546
$\mathcal{J}_{v^a}$	(624)	396	(628)	380	(931)	940
$\mathcal{J}_{Z20}$	(687)	820	(680)	678	(1058)	1129
$\mathcal{J}_{data}$	1609	2088	1622	2160	2550	4614
$\mathcal{J}$	2624	3702	2644	3810	4008	8322

(b) Standard deviations

	Year 1	Year 2	Year 3
$\mathcal{J}_F$	136 000	131 000	146 057
$\mathcal{J}_{mod}$	38	38	72
$\mathcal{J}_{data}$	60	59	45
$\mathcal{J}$	72	73	89

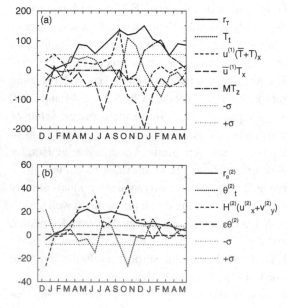

Figure 5.5.7 Time series of Year 3 term balances for the intermediate coupled model for (a) the anomalous SST equation in W m^{-2} at (135.85°W, 3.5°S) and (b) for the anomalous lower-layer thickness equation in 10^{-6} m s^{-1} at (156.38°W, 0.5°S). All are daily values, spaced thirty days apart and joined by line segments for clarity. The standard error σ in (a) is 54 W m^{-2}; in (b) it is 8 × 10^{-6} m s^{-1} (after Bennett *et al.*, 2000).

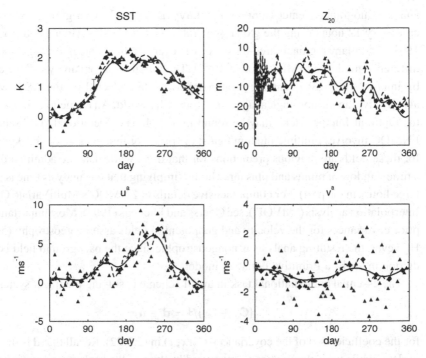

Figure 5.5.8 Time series of SST, Z20 and u^a, v^a at (180°,0°) for Year 1. Diamonds: 30-day averaged data; solid lines: corresponding inverse estimate; triangles: five-day averaged data; broken lines: corresponding inverse estimate.

and 4000 in the 18-month episode ("Year" 3). The inversion for Year 1 has been re-peated, using five-day averaged data in place of 30-day averages. There are accordingly about 15 000 of the former. A time series of the inverse SST is shown in Fig. 5.5.8. The inverse is unable to fit the data to within the 0.3° standard error of measurement, not because the data are of lower quality but because the time scales of the null hypothesis and intermediate-model dynamics are too long. The figure emphasizes that the inverse is indeed a "fixed interval smoother", and that the number M of data is not a serious restriction on the indirect representer method. Contemporary matrix manipulation tech-niques are not severely strained at $M = 10^5$ (see e.g., Egbert, 1997, §2.3; Daley and Barker, 2000).

5.6 Sampler of oceanic and atmospheric data assimilation

5.6.1 3DVAR for NWP and ocean climate models

Operational Numerical Weather Prediction relies upon timely, robust and accurate esti-mates of initial conditions. For example, the US Navy's Fleet Numerical Meteorology and Oceanography Center (www.fnmoc.navy.mil) produces a global six-day fore-cast every 12 hours using the global spectral model NOGAPS (Hogan and Rosmond, 1991), and many regional three-day forecasts every 12 hours using the regional finite-difference model COAMPS (Hodur, 1997). The initial fields or "analyses" are created by interpolating vast quantities of atmospheric data collected by the US Navy, and also by civilian meteorological agencies around the world. An elementary introduction to "Optimal Interpolation" may be found in §2.2.4. The operational "OI" scheme at FNMOC involves multiple random fields; the priors or first guesses or "backgrounds" for these fields are previous predictions for that time, while data are admitted through a time window of minus and plus three hours (implying that the analysis time is at least three hours in the past). For comprehensive details of FNMOC's MultiVariate Optimal Interpolation analysis ("MVOI"), see Goerss and Phoebus (1993). Most importantly, the prior covariances for the velocity and geopotential fields assume geostrophy (5.1.39). However, the resulting analysis is nongeostrophic since the background field is a pre-diction made by a Primitive Equation model.

The essential computational task in an OI scheme is solving the linear system

$$\{\mathbf{C}_q + \mathbf{C}_\epsilon\}\beta = \mathbf{d} - \mathbf{u}_F \tag{5.6.1}$$

for the coefficients β of the covariances $\mathbf{C}_q(x, t)$ in (2.2.22). Recall that $\mathbf{d}$ is the vector of data, while $\mathbf{u}_F$ is the vector of "measured" values of the background $u_F(x, t)$. Note that MVOI replaces the single spatial coordinate x in (2.2.22) with three spatial co-ordinates, one of which may be pressure as in (5.1.1). The dimension of the system is the number of data, which can be in excess of 10^5 in a global analysis. MVOI reduces the di-mension by analyzing the data in regions. The coefficient matrix in (5.6.1) is symmetric and positive definite, so MVOI solves the system by Cholesky factorization (Press et al., 1986, §2.9). Should small negative eigenvalues be encountered, MVOI arbitrarily increases the diagonal of $\mathbf{C}_\epsilon$, that is, it increases the variances of the measurement errors until positivity is restored.

The solution of (5.6.1) also mimizes the penalty function

$$J[\beta] = \frac{1}{2}\beta^{\mathrm{T}}\{\mathbf{C}_q + \mathbf{C}_\epsilon\}\beta - \beta^{\mathrm{T}}(\mathbf{d} - \mathbf{u}_F). \tag{5.6.2}$$

Preconditioned conjugate gradient searches (Press et al., 1986, §10.6) for the unique minimum of J may be efficiently implemented on multiprocessor computers. System dimensions of 10^5 or larger can be managed, and so regionalization is not necessary. This algorithm for implementing OI has become known as "3DVAR", and is being introduced at many NWP centers such as FNMOC (Daley and Baker, 2000), at the United Kingdom Meteorological Office (Lorenc et al., 2000) and at the NASA Global Modeling and Assimilation Office (www.polar.gsfc.nasa.gov). Oceanographers can learn much from these operational NWP centers, concerning the real-time quality-control of vast data sets and the devising of prior multivariate covariances.

A fast algorithm for the spatial "convolution" (3.1.25), essential to efficient im-plementation of physically realizable inverse models, is also required for 3DVAR.

Passi *et al.* (1996) propose a product-polynomial algorithm for convolving with the "bell-shaped" covariance (3.1.24), as an alternative to solving the pseudo-heat equation (3.1.26) subject to (3.1.27).

The need to initialize and validate climate models is stimulating major applications of "OI" to the global upper ocean. Carton *et al.* (2000a, 2000b) apply 3DOI to 45 years of upper-ocean data. Their backgrounds for the fields of temperature, salinity and current are the predictions of the standard Primitive Equation model MOM2 from the NOAA Geophysical Fluid Dynamics Laboratory (`www.gfdl.gov/~smg/ MOM/MOM.html`). Carton *et al.* pay careful attention to the mean background errors or "biases" following Dee and da Silva (1998), and also hypothesize a detailed but structurally simple multivariate covariance for the background errors.

5.6.2   4DVAR for NWP and ocean climate models

Inverse modeling often involves compromises. A common assumption is that the equations of motion are exactly correct, and that only the initial conditions and some dynamical parameters should be perturbed in order to fit the data. This would seem entirely reasonable if the smoothing or assimilation interval of interest is rather less than the evolution time scale of the dynamics: say, rather less than three days on synoptic scales in the mid-latitude troposphere, and rather less than three months on planetary scales in the tropical Pacific Ocean. There are two major variational inverse models of this kind, which are "strong constraint" assimilations in the terminology of Sasaki (1970).

The first is the "4DVAR" program in support of operational Numerical Weather Prediction at the European Centre for Medium-range Weather Forecasting (ECMWF). The project is described in a major series of papers: Rabier *et al.* (2000), Mahfouf and Rabier (2000) and Klinker *et al.* (2000). The model is a spectral representation of the global atmosphere, with about 10^7 spatial variables per time step. Variational assimilation is performed in six-hour intervals, from $t - 3$ hours to $t + 3$ hours, with vast amounts of tropospheric data being smoothed throughout the interval. There are $O(10^5)$ surface data alone. The perturbed state at time $t = 0$ hours becomes the initial condition for a forecast out to $t = 168$ hours, and the skill of the forecast is the basis for assessing the utility of the variational assimilation.

The second project is the "Estimation of the Circulation and Climate of the Ocean" (ECCO) Consortium (Stammer *et al.*, 2000). The estimation is based on the MIT nonhydrostatic General Circulation Model (Marshall *et al.*, 1997a,b). The tangent–linear and corresponding adjoint operators are constructed with a symbolic algorithm (Giering and Kaminsky, 1997), as described in Marotske *et al.* (1999). Ocean circulation is sustained by surface fluxes, both in reality and in models. These fluxes are poorly known, and so it is desirable that they should be perturbed along with the initial conditions in the search for a better fit to data. With $\frac{1}{4}^\circ$ resolution globally, there are about 2×10^8 initial variables, and about 10^7 surface fluxes per time step. The latter need not be perturbed independently at every time step. Even so, the computational challenge is clearly enormous, yet impressive progress is being made with $1°$ grids and years of data (Stammer *et al.*, 2000).

5.6.3 Correlated errors

The penalty functional (1.5.7) for controlling the toy model (1.2.6)–(1.2.8) does not contain products of $b(t)$ and $f(x, t)$, for example. The statistical interpretation (2.2.3) is the hypothesis that these boundary errors and forcing errors are uncorrelated. Bogden (2001) argues that boundary flow errors can be correlated to wind errors inside an ocean region, since both can be correlated to wind errors outside the region. Thus the hypothesis should include a nonvanishing cross-covariance:

$$C_{fb}(x, t, s) = E(f(x, t)b(s)).\qquad(5.6.3)$$

For consistency, we should change the notation for autocovariances, from (2.2.2) to

$$C_{ff}(x, t, y, s) = E(f(x, t)f(y, s)),\qquad(5.6.4)$$

for example. The estimator of maximum likelihood for such multivariate normal fields is

$$J[u] = (f\ \bullet, b\ *)\begin{bmatrix} C_{ff} & C_{fb} \\ C_{bf} & C_{bb} \end{bmatrix}^{-1}\begin{pmatrix} \bullet & f \\ * & b \end{pmatrix} + \cdots.\qquad(5.6.5)$$

The matrix inverse is defined as a matrix-valued kernel: see (1.5.11) and (1.5.12).

Exercise 5.6.1

Assume the other errors are uncorrelated with f or b or with each other, that is, assume the rest of the estimator (1.5.9) is unaltered. Derive the Euler–Lagrange equations for the penalty functional (5.6.5). Show that the weighted residual still obeys (1.3.1)–(1.3.3), but the inverse estimates for the dynamical and boundary residuals are, respectively:

$$\hat{f} = C_{ff}\bullet\lambda + C_{fb}*\lambda,\qquad(5.6.6)$$
$$\hat{b} = C_{bf}\bullet\lambda + C_{bb}*\lambda,\qquad(5.6.7)$$

where the blob products are evaluated inside the region, while the star products are evaluated on the boundary.   □

5.6.4 Parameter estimation

The constant phase speed c in the toy model (1.1.1) has been kept fixed up to now. Yet the fit to data may be improved by varying c. To this end, the penalty functional $J[u, c]$ in (1.5.9) may be augmented (Bennett, 1992, §10.2):

$$\mathcal{K}[u, c] = \sigma_f^{-2}(c - c_0)^2 + J[u, c],\qquad(5.6.8)$$

where c_0 is a prior for c, and σ_c^2 is the hypothesized variance of the prior error. Varying $\mathcal{K}$ with respect to c and $u(x, t)$ yields the extremal condition

$$\hat{c} = c_0 + \sigma_c^2 \frac{\partial \hat{u}}{\partial x} \bullet \lambda \qquad (5.6.9)$$

and the Euler–Lagrange equations for $\mathcal{J}$ as before. Note that $\hat{u}$ and λ depend upon $\hat{c}$, so (5.6.9) is highly nonlinear even though the dynamics of the toy model are linear. However, iteration schemes for solving (5.6.9) are readily devised (Eknes and Evensen, 1997, who consider linear Ekman layer dynamics with an unknown eddy viscosity), and interlaced with iterated representer algorithms in the case of nonlinear dynamics (Muccino and Bennett, 2002, who consider the Korteveg–DeVries equation with unknown parameters for phase speed, amplitude and dispersion). Convergence of these schemes is in no way assured.

5.6.5   Monte Carlo smoothing and filtering

Consider the toy nonlinear model (3.3.1)–(3.3.3), where the random inputs f, i and b have covariances C_f, C_i and C_b respectively. The methods of §3.2.4 may be used to generate pseudo-random samples of the inputs consistent with their respective covariances. A pseudo-random sample of the state $u(x, t)$ is then obtained by integrating (3.3.1)–(3.3.3). Sample estimates of the expectation $Eu(x, t)$ and covariance $C_u(x, t, y, s)$ follow from repeated generation and integration.

The sample moments of u may then be used for space–time optimal interpolation of data collected in some time interval $0 < t < T$, as outlined in §2.2.4. The prior for the OI or best linear unbiased estimate (2.2.22) would not be $u_F(x, t)$, but rather the sample estimate of $Eu(x, t)$, while the covariance $C_q(x, t, y, s)$ would be the sample estimate of $C_u(x, t, y, s)$. As a consequence of the nonlinearity of (3.3.1), the OI estimate is not an extremum of the penalty functional (1.5.9), even if W_f were related to C_f through (1.5.11), (1.5.12), etc. That is, the OI estimate is not the solution of an inverse model. Nevertheless the attraction of such "Monte Carlo smoothing" is obvious: there is no need to linearize the dynamics, nor is it necessary to derive the adjoint dynamics.

Storing $C_u(x, t, x_m, t_m)$, where (x_m, t_m) is a data point, may not be feasible for all x, t and for $1 < m < M$, but it may be feasible to store $C_u(x, t_m, y, t_m)$, for all x, y and for one time t_m. Data collected at the time t_m may thus be optimally interpolated in space, provided it is assumed that the data errors are uncorrelated in time. This Monte Carlo filtering method has become known as the "Ensemble Kalman Filter" or EnKF (Evensen, 1994). For its application to operational forecasting of the North Atlantic Ocean, see http://diadem.nersc.no/project; for application to seasonal-to-interannual forecasting of the Tropical Pacific Ocean, see Keppenne (2000). For a careful comparison of the computational efficiency of the EnKF with that of the indirect iterated representer algorithm, in the context of an hydrological model and satellite observations of soil moisture, see Reichle et al. (2001, 2002).

Exercise 5.6.2

Why is it necessary in the EnKF to assume that the errors in data collected at different times are uncorrelated?   □

5.6.6   Documentation

The most important but most neglected aspect of ocean modeling is documentation. It is no exaggeration to state that if a model has not been documented, it does not exist. Documentation is, of course, a tedious chore and the scarce resources for academic research rarely support anything so pedestrian. One outstanding exception is the Modular Ocean Model version 3, or "MOM3" (www.gfdl.gov/~smg/MOM/MOM.html). Inverse models are vastly more complex (see for example Figs. 3.1.1 and 3.1.2), so their documentation is even more important and even more neglected.

Ide *et al.* (1997) have made the following reasonable proposal of a standard notation for data assimilation. All models are eventually subject to numerical approximation of one form or another, so in computational practice the state **x** is a vector of finite dimension I:

$$\mathbf{x} = (x_1, x_2, \ldots, x_I). \tag{5.6.10}$$

The value of I is the product of the number of fluid variables (velocity components, temperature, pressure, etc.) and the number of computational degrees of freedom in space (the number of grid points, in a finite-difference model). The model evolution in a single time step is

$$\mathbf{x}_n^b = \mathbf{M}_n \left[\mathbf{x}_{n-1}^b \right], \tag{5.6.11}$$

where $\mathbf{M}_n$ is in general a nonlinear operator, and the initial vector is $\mathbf{x}_0^b$. The subscript n in (5.6.11) indicates the state vector or nonlinear operator at time t_n. The superscript b indicates that the vector will be the background for an optimal estimate of the state. Thus $\mathbf{x}_n^b$, $(0 \leq n \leq N)$ lumps the field $u_F(x, t)$, $(0 < x < L,\ 0 \leq t \leq T)$ of §1.1.1. Observations are taken at selected times t_{n_j}; at any such time, these data comprise a vector of dimension K:

$$\mathbf{y}_j^o = \left(y_{j1}^o, y_{j2}^o, \ldots, y_{jK}^o \right). \tag{5.6.12}$$

The measurement functional $\mathbf{H}_k$ is in general nonlinear;

$$\mathbf{y}_j^b \equiv \mathbf{H}_j \left[\mathbf{x}_{n_j}^b \right], \tag{5.6.13}$$

for example, is the measured value of the background at time t_{n_j}. The stage is now set for defining an inverse model in terms of a weighted least-squares penalty function (Uboldi and Kamachi, 2000), analogous to (1.5.9).

The standard notation proposed by Ide *et al.* (1997) is becoming widely accepted. This aids high-level dialogue, at the expense of insight into dynamical detail. For example, it is not obvious from the abstract finite-dimensional equation (5.6.11) that the tangent linearization of the evolution operator M_n may be unphysical: see §3.3.4. Nor is it obvious that spatial and temporal irregularity arising from unsuitable weighting (see §2.6), or from ill-posedness of the forward model (see Chapter 6), are fundamental issues.

Documentation will remain a crucial challenge, even with the advent of standard notations such as that of Ide *et al.* (1997). Preliminary documentation exists (Chua and Bennett, 2001) for IOM (Inverse Ocean Model), a modular code for the iterated, indirect representer algorithm of Fig 3.1.2. The software engineering for such modular systems must accommodate a wide range of models and modeling practices, and yet retain high computational performance. An elegant graphical representation for the Project d'Assimilation par Logiciel Multi-méthodes ("PALM"), a universal coupler of models and data, has been devised by Lagarde *et al.* (2001) as an aid to software development.

Chapter 6

Ill-posed forecasting problems

The "toy" forward model introduced in Chapter 1 defines a well-posed mixed initial value–boundary value problem. The associated operator (wave operator plus initial operator plus boundary operator) is invertible, or nonsingular. Specifying additional data in the interior of the model domain renders the problem overdetermined. The operator becomes uninvertible, or singular. The difficulty may be resolved by constructing the generalized inverse of the operator, in a weighted least-squares sense.

An important class of regional models of the ocean or atmosphere defines an ill-posed initial-boundary value problem, regardless of the choice of open boundary conditions. All flow variables may as well be specified on the open boundaries. The excess of information may be regarded as data on a bounding curve, rather than at an interior point. The difficulty may again be resolved by constructing the generalized inverse in the weighted least-squares sense. The Euler–Lagrange equations form a well-posed boundary value problem in space–time. Solving them by forward and backward integrations is precluded, since no partitioning of the variational boundary conditions yields well-posed integrations. The penalty functional must be minimized directly.

Things are different if the open region is moving with the flow.

6.1 The theory of Oliger and Sundström

We have been assuming that our forward model constitutes a well-posed problem. That is, just sufficient information is given about the forcing $F(x, t)$, the boundary values $B(t)$ and initial values $I(x)$ in order to ensure the existence of a solution that is unique for each choice of F, B and I, and which depends continuously upon smooth changes to F, B and I. For linear models, the proof of uniqueness also implies continuous dependence, and even implies existence (Courant and Hilbert, 1962, Ch. VI, §10). Sometimes we are able to establish existence (and the other requirements) by displaying the general solution explicitly. If the explicit solution has been obtained after making an assumption about the solution, such as the variables being separable, then uniqueness must be established first. Uniqueness and continuous dependence can be established for some nonlinear models, such as the shallow-water equations (Oliger and Sundström, 1978), but existence is usually an open question. Existence has been established for inviscid and viscous incompressible flow in the plane (Ladyzhenskaya, 1969). These results have been extended to certain quasigeostrophic flows (Bennett and Kloeden, 1981).

In this section we shall briefly review the uniqueness of solutions of some simple linear models, and then use these ideas to determine the number of boundary conditions needed at open boundaries. Following Oliger and Sundström (1978), it will be shown that there is a difficulty with the Primitive Equations, but it will be argued that the ill-posedness can be resolved by generalized inversion of the open-ocean Primitive Equation model. In subsequent sections we shall address the special methods for finding the generalized inverse, given that the forward model is ill-posed.

6.2   Open boundary conditions for the linear shallow-water equations

In §1.1.2 we verified that the initial-boundary value problem for the one-dimensional linear wave equation of §1.1.1 has a unique solution. Now consider a linear shallow-water model:

$$\frac{\partial \mathbf{u}}{\partial t} = -g \nabla h, \tag{6.2.1}$$

$$\frac{\partial h}{\partial t} = -H \nabla \cdot \mathbf{u}, \tag{6.2.2}$$

where $\mathbf{u} = \mathbf{u}(\mathbf{x}, t)$ is a planar velocity field, $\mathbf{x}$ is a point in the plane, t is time, $h = h(\mathbf{x}, t)$ is the sea-level disturbance, g is the gravitational acceleration, H is the mean depth, and $\nabla = (\frac{\partial}{\partial x}, \frac{\partial}{\partial y})$. Suitable initial conditions are

$$\mathbf{u}(\mathbf{x}, 0) = \mathbf{0}, \tag{6.2.3}$$

$$h(\mathbf{x}, 0) = 0. \tag{6.2.4}$$

We need not include forcing in (6.2.1) or (6.2.2), nor nonzero initial values in (6.2.3) and (6.2.4), since we are interested in the difference between two solutions having the same forcing, initial values and boundary values.

It follows from (6.2.1) and (6.2.2) that

$$\frac{d}{dt} \iint_{\mathcal{D}} \left(\frac{1}{2} H |\mathbf{u}|^2 + \frac{1}{2} gh^2 \right) da = -gH \int_{\mathcal{B}} h\mathbf{u} \cdot \hat{\mathbf{n}} \, ds, \qquad (6.2.5)$$

where $\mathcal{D}$ is the spatial domain, $\mathcal{B}$ is its boundary, da is an area element in $\mathcal{D}$, ds is an arc element on $\mathcal{B}$, and $\hat{\mathbf{n}}$ is an outward normal on $\mathcal{B}$. It is clear from (6.2.3) and (6.2.4) that the area integral on the lhs of (6.2.5) vanishes at $t = 0$, so if the rhs is nonpositive for all $t \geq 0$, then the area integral vanishes for all $t \geq 0$; hence $\mathbf{u}(\mathbf{x}, t) = \mathbf{0}$ and $h(\mathbf{x}, t) = 0$. Uniqueness would then be established. That is,

$$h\mathbf{u} \cdot \hat{\mathbf{n}} \geq 0 \qquad (6.2.6)$$

on $\mathcal{B}$, for all $t \geq 0$, would ensure uniqueness. For example:

(i) specify $\mathbf{u} \cdot \hat{\mathbf{n}} = 0$ on $\mathcal{B}$

or

(ii) specify $h = 0$ on $\mathcal{B}$

or

(iii) specify $\mathbf{u} \cdot \hat{\mathbf{n}} = \alpha \sqrt{\frac{g}{H}} h$ on $\mathcal{B}$,

where α is a positive constant.

Notice that one of (i), (ii) or (iii) would suffice for uniqueness; there is no need to specify both the normal velocity and the sea-level elevation. That would overdetermine the solution.

Exercise 6.2.1

Consider the difference between two solutions of the linear shallow-water equations, corresponding to two different sets of forcing, initial and boundary values. Show that the total energy of the difference is controlled by the differences in the inputs. Hint: let

$$\| \mathbf{F} \|(t) \equiv \left(\iint_{\mathcal{D}} |\mathbf{F}(\mathbf{x}, t)|^2 \, da \right)^{\frac{1}{2}} ;$$

it may be shown that

$$\left| \iint_{\mathcal{D}} \mathbf{F} \cdot \mathbf{v} \, da \right| \leq \| \mathbf{F} \| \cdot \| \mathbf{v} \| .$$

□

6.3   Advection: subcritical and supercritical, inflow and outflow

In a small step towards nonlinearity, let us now include a constant advecting velocity $\mathbf{U}$ in the linear shallow-water equations:

$$\frac{\partial \mathbf{u}}{\partial t} + \mathbf{U} \cdot \nabla \mathbf{u} = -g\nabla h, \qquad (6.3.1)$$

$$\frac{\partial h}{\partial t} + \mathbf{U} \cdot \nabla h = -H\nabla \cdot \mathbf{u}. \qquad (6.3.2)$$

Hence

$$\frac{dE}{dt} \equiv \frac{d}{dt} \iint_{\mathcal{D}} \frac{1}{2}(H|\mathbf{u}|^2 + gh^2)\, da$$

$$= -\int_{\mathcal{B}} \left\{ \frac{1}{2}(H|\mathbf{u}|^2 + gh^2)\mathbf{U} \cdot \hat{\mathbf{n}} + gHh\mathbf{u} \cdot \hat{\mathbf{n}} \right\} ds. \qquad (6.3.3)$$

In order to establish uniqueness, we must arrange for the rhs of (6.3.3) to be nonpositive. The integrand in (6.3.3) is a quadratic form:

$$\{\ \} = \frac{1}{2}H(\mathbf{u}^{\mathrm{T}}, \sigma h) \begin{pmatrix} \mathbf{I} & c\hat{\mathbf{n}} \\ c\hat{\mathbf{n}}^{\mathrm{T}} & U \end{pmatrix} \begin{pmatrix} \mathbf{u} \\ \sigma h \end{pmatrix}, \qquad (6.3.4)$$

where $\sigma = (g/H)^{\frac{1}{2}}$, $U = \mathbf{U} \cdot \hat{\mathbf{n}}$, $\mathbf{I}$ is the 2×2 unit matrix and $c = (gH)^{\frac{1}{2}}$. We must determine if the 3×3 matrix in (6.3.4) is definite or indefinite. Its eigenvalues λ satisfy

$$\begin{vmatrix} U - \lambda & 0 & c\hat{n}_1 \\ 0 & U - \lambda & c\hat{n}_2 \\ c\hat{n}_1 & c\hat{n}_2 & U - \lambda \end{vmatrix} = 0, \qquad (6.3.5)$$

that is,

$$(U - \lambda)\{(U - \lambda)^2 - c^2\} = 0.$$

Hence the three eigenvalues are

$$\lambda_0 = U, \quad \lambda_\pm = U \pm c.$$

Note that these are defined at each point on the boundary, and that $U \equiv \mathbf{U} \cdot \hat{\mathbf{n}}$ is the local value.

There are several cases to consider:

I. OUTFLOW: $U > 0$

  (a) SUPERCRITICAL: $U > c > 0$

All three eigenvalues are positive. So, regardless of the boundary values, $\frac{dE}{dt} \le 0$ and uniqueness follows with

NO BOUNDARY CONDITIONS.

(b) SUBCRITICAL: $0 < U < c$

Then $\lambda_0 = U > 0$, $\lambda_+ = U + c > 0$, $\lambda_- = U - c < 0$. Hence $\frac{dE}{dt} \leq 0$
provided $(\mathbf{u}^T, \sigma h)$ is orthogonal to the eigenvector $\boldsymbol{\mu}_-$ associated with λ_-.
This may be assured with

ONE BOUNDARY CONDITION.

In this case $U - \lambda_- = U - (U - c) = c$, so $\boldsymbol{\mu}_- = (\mu, v, \xi)^T$ satisfies

$$\begin{pmatrix} c & 0 & c\hat{n}_1 \\ 0 & c & c\hat{n}_2 \\ c\hat{n}_1 & c\hat{n}_2 & c \end{pmatrix} \begin{pmatrix} \mu \\ v \\ \xi \end{pmatrix} = \mathbf{0}. \tag{6.3.6}$$

A non-normalized choice is

$$\begin{pmatrix} \mu \\ v \\ \xi \end{pmatrix} = \begin{pmatrix} -\hat{n}_1 \\ -\hat{n}_2 \\ 1 \end{pmatrix}, \tag{6.3.7}$$

hence we require

$$(u, v, \sigma h) \begin{pmatrix} -\hat{n}_1 \\ -\hat{n}_2 \\ 1 \end{pmatrix} = 0, \tag{6.3.8}$$

or

$$\boxed{\mathbf{u} \cdot \hat{\mathbf{n}} = \sigma h} \tag{6.3.9}$$

II. INFLOW: $U < 0$

(a) SUPERCRITICAL: $0 < c < |U|$
   All three eigenvalues are negative so we need

THREE BOUNDARY CONDITIONS:

$$\mathbf{u} = \mathbf{0}, \tag{6.3.10}$$

$$h = 0. \tag{6.3.11}$$

(b) SUBCRITICAL: $0 < |U| < c$
   Hence $\lambda_0 = U < 0$, $\lambda_+ = U + c > 0$, $\lambda_- = U - c < 0$: two eigenvalues
   are negative, so we need

TWO BOUNDARY CONDITIONS.

Exercise 6.3.1

Find $\boldsymbol{\mu}_0$, $\boldsymbol{\mu}_-$.   □

The above results may be summarized:

		INFLOW $(U < 0)$	OUTFLOW $(U > 0)$
SUBCRITICAL	$(|U| < c)$	2	1
SUPERCRITICAL	$(c < |U|)$	3	0

Again, the situation will in general vary around the boundary.

Exercise 6.3.2
Consider the cases $U = 0$, $|U| = c$. ☐

We conclude that it is possible to determine the correct number of boundary conditions for a linearized shallow-water model. That is, we may determine the number that ensures the uniqueness of solutions, and hence continuous dependence upon the inputs.

6.4   The linearized Primitive Equations in isopycnal coordinates: expansion into internal modes; ill-posed forward models with open boundaries

The Primitive Equations were presented in §5.1 in isobaric or "pressure" coordinates. Let's denote these coordinates by (x', y', p, t'). Now consider isopycnal or "density" coordinates (x, y, α, t), where α is the specific volume or inverse density: $\alpha \equiv \rho^{-1}$. Define the Montgomery potential m by

$$m = \alpha p + \phi,$$

where $\phi = gz$ is the geopotential.

Exercise 6.4.1
Show that

$$\frac{\partial \mathbf{u}}{\partial t} + \mathbf{u} \cdot \nabla \mathbf{u} = -\nabla m, \tag{6.4.1}$$

$$\frac{\partial m}{\partial \alpha} = p, \tag{6.4.2}$$

$$\frac{\partial^2 p}{\partial t \partial \alpha} + \nabla \cdot \left(\mathbf{u} \frac{\partial p}{\partial \alpha} \right) = 0. \tag{6.4.3}$$

Explain the meaning of the partial derivatives. Note that there is no "diapycnal" advection in (6.4.1) and (6.4.3). This is a consequence of

(1) the combined first and second laws of thermodynamics:

$$T d\eta = de + p d\alpha, \tag{6.4.4}$$

where T is temperature, η is entropy, and e is internal energy;

(2) the assumption of isentropic motion;

(3) the assumption that the internal energy of the fluid is constant.

The isopycnal form of the Primitive Equations need not be so restrictive, but this form suffices for our immediate purpose. Linearize about a uniform horizontal velocity U, a static-state Montgomery potential $m = M(\alpha)$, and a pressure $p = P(\alpha)$, where $P = \frac{dM}{d\alpha}$:

$$\frac{\partial \mathbf{u}}{\partial t} + \mathbf{U} \cdot \nabla \mathbf{u} = -\nabla m, \tag{6.4.5}$$

$$\frac{\partial m}{\partial \alpha} = p, \tag{6.4.6}$$

$$\frac{\partial^2 p}{\partial t \partial \alpha} + \mathbf{U} \cdot \nabla \frac{\partial p}{\partial \alpha} + \frac{dP}{d\alpha} \nabla \cdot \mathbf{u} = 0. \tag{6.4.7}$$

The fields $\mathbf{u}$, m and p are now perturbations about $\mathbf{U}$, M and P. Separate variables according to

$$\mathbf{u}(\mathbf{x}, \alpha, t) = \mathbf{u}'(\mathbf{x}, t) A(\alpha), \tag{6.4.8}$$

$$m(\mathbf{x}, \alpha, t) = m'(\mathbf{x}, t) A(\alpha), \tag{6.4.9}$$

$$p(\mathbf{x}, \alpha, t) = p'(\mathbf{x}, t) \frac{dA}{d\alpha}(\alpha). \tag{6.4.10}$$

Derive the separated equations

$$\frac{\partial \mathbf{u}'}{\partial t} + \mathbf{U} \cdot \nabla \mathbf{u}' = -\nabla m', \tag{6.4.11}$$

$$\frac{\partial m'}{\partial t} + \mathbf{U} \cdot \nabla m' + c^2 \nabla \cdot \mathbf{u}' = 0, \tag{6.4.12}$$

and

$$\frac{d^2 A}{d\alpha^2} - c^{-2} \left(\frac{dP}{d\alpha} \right) A = 0, \tag{6.4.13}$$

where the separation constant c^2 has the dimensions of (speed)^2.   □

The unseparated ocean boundary conditions are:

$$p\left(\mathbf{x}, \alpha^{(a)}, t\right) = p^{(a)}(\mathbf{x}, t), \tag{6.4.14}$$

where $\alpha^{(a)}$ is an isopycnal surface in contact with the atmosphere, which is at pressure $p^{(a)}$;

$$m^{(b)}(\mathbf{x}, t) \equiv m\left(\mathbf{x}, \alpha^{(b)}, t\right) = \alpha^{(b)} p^{(b)} + \phi^{(b)}, \tag{6.4.15}$$

assuming that the ocean bottom at

$$z = z^{(b)}(\mathbf{x}) = \phi^{(b)}/g \tag{6.4.16}$$

is the isopycnal surface $\alpha = \alpha^{(b)}$. Hence the separated boundary conditions for perturbations are:

$$\frac{dA}{d\alpha} = 0 \qquad (6.4.17)$$

at $\alpha = \alpha^{(a)}$;

$$A = \alpha \frac{dA}{d\alpha} \qquad (6.4.18)$$

at $\alpha = \alpha^{(b)}$.

The system (6.4.11), (6.4.12) is the linearized shallow water equations with phase speed c. The system (6.4.13), (6.4.17), and (6.4.18) comprises a regular Sturm–Liouville problem (Stakgold, 1979), with eigenvalues $c_0 > c_1 > \cdots > c_n > \cdots > 0$ and eigenmodes $A_0(\alpha), A_1(\alpha), \ldots, A_n(\alpha), \ldots$.

Exercise 6.4.2

Show that

$$A^{(b)^2} = \alpha(b) \int_a^b \left\{ \left(\frac{dA}{d\alpha}\right)^2 + c^{-2} \frac{dP}{d\alpha} A^2 \right\} d\alpha, \qquad (6.4.19)$$

where $A^{(b)} = A(\alpha^{(b)})$. Note that $\frac{dP}{d\alpha} < 0$. Hence the external mode, for which A is approximately independent of α, has phase speed c_0 satisfying

$$c_0^2 \cong \alpha^{(b)} \left(P^{(b)} - P^{(a)} \right). \qquad (6.4.20)$$

□

The internal modes have lower phase speeds:

$$c_n^2 \cong c_0^2 \frac{\left(\alpha^{(a)} - \alpha^{(b)}\right)}{\frac{1}{2}\left(\alpha^{(a)} + \alpha^{(b)}\right)} J_n^{-2}, \qquad (6.4.21)$$

where $J_n = 0(n)$ as $n \to \infty$. For the Southern Ocean,

$$c_0 \cong 220 \text{ m s}^{-1}, \; c_1 = 1 \text{ m s}^{-1}, \; c_2 = 0.5 \text{ m s}^{-1}, \; c_3 \cong 0.3 \text{ m s}^{-1}, c_4 = 0.2 \text{ m s}^{-1}. \qquad (6.4.22)$$

In the Antarctic Circumpolar Current,

$$|\mathbf{u}| \simeq 0.6 \text{ m s}^{-1}. \qquad (6.4.23)$$

Hence

$$c_0 > c_1 > |\mathbf{u}| > c_2 > c_3 \ldots. \qquad (6.4.24)$$

Consider the two lowest modes: $n = 0, 1$. The amplitudes $\mathbf{u}_n(\mathbf{x}, t)'$, $m_n(\mathbf{x}, t)'$ satisfy the linearized shallow water equations (6.4.11), (6.4.12). The flow is everywhere subcritical so two boundary conditions are needed at inflow, and one at outflow.

For all other modes ($n = 2, 3, \ldots$), the flow is in general supercritical (wherever $|\mathbf{U} \cdot \hat{\mathbf{n}}| > c_2$), so three boundary conditions are needed at inflow, while none is needed at outflow. The problem is that we don't usually integrate the Primitive Equations mode by mode; we usually specify boundary conditions at each level, or each value of α. Suppose we choose two BCs at inflow and one BC at outflow. Then modes 2, 3, ... are underspecified at inflow and overspecified at outflow. Underspecification is often incorrectly eliminated by the imposition of what we intended to be computational boundary conditions; in these circumstances they acquire a dynamical role. Overspecification leads to computational noise that is usually suppressed by smoothing the fields. The ill-posedness of the open boundary problem for the Primitive Equations cannot be *solved*, but it can be *resolved* by generalized inversion.

6.5   Resolving the ill-posedness by generalized inversion

To demonstrate the approach, it suffices to consider the shallow-water equations (Bennett, 1992; Bennett & Chua, 1994). Prescribe three boundary conditions on the open boundary. This overdetermines the problem, so do not seek an exact solution of the equations of motion. Rather, seek a weighted, least-squares best-fit to all the information. The dynamics are thus

$$\frac{\partial \mathbf{u}}{\partial t} + \mathbf{U} \cdot \nabla \mathbf{u} = -c\nabla q + \mu, \tag{6.5.1}$$

$$\frac{\partial q}{\partial t} + \mathbf{U} \cdot \nabla q = -c\nabla \cdot \mathbf{u} + \chi, \tag{6.5.2}$$

where $q = m/c$, while μ and χ are misfits or residuals. A simple penalty functional is

$$\mathcal{J}[\mathbf{u}, q] = W_D \iint_{\mathcal{D}} da \int_0^T dt\, (|\mu|^2 + \chi^2)$$

$$+ W_B \int_{\mathcal{B}} ds \int_0^T dt\, \{|\mathbf{u} - \mathbf{u}_B|^2 + (q - q_B)^2\}$$

$$+ W_I \iint_{\mathcal{D}} da\, \{|\mathbf{u} - \mathbf{u}_I|^2 + (q - q_I)^2\} \tag{6.5.3}$$

$$+ \text{(data penalties)},$$

where $\mathcal{D}$ is the domain, $\mathcal{B}$ is the entirely open boundary, $(\mathbf{u}_B, q_B)$ are the boundary values, and $(\mathbf{u}_I, q_I)$ are the initial values. The weighting is simple in the extreme, but clarity is the issue here. A more realistic "nondiagonal" weighting should be used in practice. Note that no boundary conditions are needed at a supercritical outflow boundary, so the boundary values there are like data prescribed continuously on curves.

Exercise 6.5.1

Show that the Euler–Lagrange equations for local extrema of $\mathcal{J}$ are:

$$-\frac{\partial \mu}{\partial t} - \mathbf{U} \cdot \nabla \mu - c\nabla \chi + \cdots = \mathbf{0}, \qquad (6.5.4)$$

$$-\frac{\partial \chi}{\partial t} - \mathbf{U} \cdot \nabla \chi - c\nabla \cdot \mu + \cdots = 0, \qquad (6.5.5)$$

$$\mu = \mathbf{0}, \quad \chi = 0 \qquad (6.5.6)$$

at $t = T$,

$$-W_{\mathrm{D}}\mu + W_{\mathrm{I}}(\mathbf{u} - \mathbf{u}_{\mathrm{I}}) = \mathbf{0}, \qquad (6.5.7)$$

$$-W_{\mathrm{D}}\chi + W_{\mathrm{I}}(q - q_{\mathrm{I}}) = 0, \qquad (6.5.8)$$

both at $t = 0$, and

$$W_{\mathrm{D}}\mathbf{U} \cdot \hat{\mathbf{n}}\mu + W_{\mathrm{B}}(\mathbf{u} - \mathbf{u}_{\mathrm{B}}) + W_{\mathrm{B}}c\hat{\mathbf{n}}\chi = \mathbf{0}, \qquad (6.5.9)$$

$$W_{\mathrm{D}}\mathbf{U} \cdot \hat{\mathbf{n}}\chi + W_{\mathrm{B}}(q - q_{\mathrm{B}}) + W_{\mathrm{B}}c\hat{\mathbf{n}} \cdot \mu = 0, \qquad (6.5.10)$$

both on $\mathcal{B}$. The ellipsis $(\cdots)$ in (6.5.4) and (6.5.5) denotes data terms. The system (6.5.1), (6.5.2), (6.5.4)–(6.5.10) is a boundary value problem in the space–time domain $\mathcal{D} \times [0, T]$. □

Exercise 6.5.2

Show that if $(\mathbf{u}, q)$ and (μ, χ) satisfy the Euler–Lagrange equations (6.5.4)–(6.5.10), then

$$W_{\mathrm{D}} \iint_{\mathcal{D}} da \int_0^T dt \, (|\mu|^2 + \chi^2) + W_{\mathrm{B}} \int_{\mathcal{B}} ds \int_0^T dt \, (|\mathbf{u}|^2 + q^2)$$

$$+ W_{\mathrm{I}} \iint_{\mathcal{D}} du \, (|\mathbf{u}|^2 + q^2)_{t=0} + (\cdots)^2 = 0, \qquad (6.5.11)$$

where $(\cdots)^2$ denotes a nonnegative contribution from the data sites.

It follows immediately from (6.5.11) that

$$\mu \equiv \mathbf{0}, \quad \chi \equiv 0 \qquad (6.5.12)$$

in $\mathcal{D} \times [0, T]$,

$$\mathbf{u} = \mathbf{0}, \quad q = 0 \qquad (6.5.13)$$

in $\mathcal{B} \times [0, T]$, and

$$u = 0, \quad q = 0 \qquad (6.5.14)$$

in $\mathcal{D} \times \{0\}$. The "forward problem" for (u, q) is (6.5.1), (6.5.2), which together with (6.5.12)–(6.5.14) is overdetermined, but the solution is without question

$$u \equiv 0, \quad q = 0. \qquad (6.5.15)$$

It may be concluded that the Euler–Lagrange equations form a well-posed boundary problem in $\mathcal{D} \times [0, T]$, even though the original forward model is ill-posed. The challenge is to find a solution algorithm for the inverse, when the forward model is ill-posed. Backward and forward integrations must be avoided. □

Exercise 6.5.3

It may be difficult to accept that generalized inversion can be well-posed, even though the forward model is ill-posed. Consider a simple example, defined by an ordinary differential equation:

$$\frac{dx}{dt} = 1 \qquad (6.5.16)$$

for $0 \le t \le 1$, subject to

$$x(0) = 0, \quad x(1) = 2. \qquad (6.5.17)$$

This overdetermined problem does not have a continuous solution. Now seek a least-squares best-fit to (6.5.16) and (6.5.17), with a penalty functional

$$\mathcal{J}[x] = \int_0^1 \left(\frac{dx}{dt} - 1 \right)^2 dt + x(0)^2 + (x(1) - 2)^2. \qquad (6.5.18)$$

Derive the Euler–Lagrange equations for the best fit, and verify that they have the unique, continuous solution

$$\hat{x}(t) = \frac{1 + 4t}{3}. \qquad (6.5.19)$$

□

6.6   State space optimization

Bennett and Chua (1994) used simulated annealing and HMC to resolve the ill-posedness of an idealized, regional shallow-water model. The ill-posedness arose from specifying too much data at the open boundaries, thereby mimicking the situation that is inevitable for regional Primitive Equation models. As argued in §6.5, the ill-posedness is resolved by reformulation as a generalized inverse problem, since the associated Euler–Lagrange (EL) equations form a well-posed boundary value problem.

The EL equations are efficiently solved using representer methods, but these require backward and forward integrations. When the dynamics are shallow-water, it is always possible to partition the EL boundary conditions so that both the backward integration and the forward integration are well-posed. Such a partitioning is not possible for Primitive Equation dynamics. So, in the interest of developing generalized inversion techniques alternative to solving the EL equations, Bennett and Chua (1994) minimized the nonlinear shallow-water penalty functional by direct attack with simulated annealing and HMC. Numerical experiments with synthetic data supported the argument that the inverse is well-posed. It was also shown that assimilation of "accurate" synthetic interior data compensated for "inaccurate" synthetic boundary data. The time dependent calculations involved about 740 000 computational variables on a space–time grid. The annealing process was computed using a CM-200 Connection Machine, and animated with a frame buffer. Local extrema in the penalty functional were seen to be associated with jagged "annealing flaws" in the circulation fields. A careful annealing strategy led eventually to the global minimum and the correct, smooth circulation. The annealing samples in the final stages yielded crude posterior error statistics. An HMC approach led directly to the global minimum but provided no error statistics. Those could be obtained, albeit crudely, by importance-sampling near the minimum. HMC and other gradient methods become impractical for the estimation of smooth fields involving more than 10^6 gridded variables. In comparison, iterated representer methods have been successfully applied to the estimation of as many as 10^9 gridded variables modeling moderately nonlinear flows, such as global weather, and climatic variability of the ocean–atmosphere: see Chapter 5. We do not, however, have efficient data assimilation techniques for highly turbulent flows marked by sharp fronts, outcrops or other near-discontinuities.

Exercise 6.6.1
Construct a quartic polynomial $J = J(u)$ having a graph like a "lopsided letter w". Minimize J by simulated annealing. $\square$

6.7 Well-posedness in comoving domains

Oliger and Sundström (1978) established the uniqueness of solutions of the nonlinear shallow-water equations:

$$\frac{\partial \mathbf{u}}{\partial t} + \mathbf{u} \cdot \nabla \mathbf{u} = -g\nabla h + \mathbf{F}, \tag{6.7.1}$$

$$\frac{\partial h}{\partial t} + \mathbf{u} \cdot \nabla h + h\nabla \cdot \mathbf{u} = 0, \tag{6.7.2}$$

subject to the initial conditions

$$\mathbf{u}(\mathbf{x}, 0) = \mathbf{u}_I(\mathbf{x}), \quad h(x, 0) = h_I(\mathbf{x}).\tag{6.7.3}$$

Let $\mathbf{u}^{(1)}$, $h^{(1)}$ and $\mathbf{u}^{(2)}$, $h^{(2)}$ be two solutions for the same forcing $\mathbf{F}$ and initial values $\mathbf{u}_I$, h_I; let $\mathbf{u}' = \mathbf{u}^{(1)} - \mathbf{u}^{(2)}$ and $h' = h^{(1)} - h^{(2)}$ be the difference fields, and let

$$m = \frac{1}{2} h^{(1)} \mathbf{u}' \cdot \mathbf{u}' + \frac{1}{2} g(h')^2\tag{6.7.4}$$

be the total mechanical energy in the difference fields. Oliger and Sundström (1978) showed that

$$\frac{\partial m}{\partial t} + \nabla \cdot \left[\mathbf{u}^{(1)} m + g h^{(1)} h' \mathbf{u}' \right]$$

$$= g h' \mathbf{u}' \cdot \nabla h^{(1)} - g h' \mathbf{u}' \cdot \nabla h^{(2)} - g (h')^2 \nabla \cdot \mathbf{u}^{(2)}$$

$$+ \frac{1}{2} g (h')^2 \nabla \cdot \mathbf{u}^{(1)} - h^{(1)} \mathbf{u}' \cdot (\mathbf{u}' \cdot \nabla) \mathbf{u}^{(2)}.\tag{6.7.5}$$

It follows that

$$\frac{dE}{dt} + A \le BE,\tag{6.7.6}$$

where

$$E = E(t) = \iint_{\mathcal{D}} m d\mathbf{x},\tag{6.7.7}$$

$$A = A(t) = \int_{B} \left\{ \mathbf{u}^{(1)} \cdot \hat{\mathbf{n}} m + g h^{(1)} h' \mathbf{u}' \cdot \hat{\mathbf{n}} \right\} ds,\tag{6.7.8}$$

$$B = \left(\frac{11}{2} \right) \max_{i,\mathbf{x},t} \left\{ \left| \nabla \cdot \mathbf{u}^{(1)} \right|, \ (g/h^{(i)})^{\frac{1}{2}} \left| \nabla h^{(i)} \right| \right\}\tag{6.7.9}$$

and $\hat{\mathbf{n}}$ is the outward unit normal on the boundary B of the *fixed* domain $\mathcal{D}$. Integrating (6.7.6) yields

$$E(t) \le E(0) - \int_{0}^{t} e^{B(t-r)} A(r) dr.\tag{6.7.10}$$

Note that the inverse timescale B is a constant. Now $E(0) = 0$ since the two solutions satisfy the same initial conditions. It has been shown in §3.3.4 that $h^{(1)} > 0$, provided $h_I > 0$. It is therefore clear that any boundary conditions ensuring $A \ge 0$ also ensure $E(t) \equiv 0$, that is, the mixed-initial-boundary value problem has a unique solution.

Exercise 6.7.1

It was shown in §6.2 that certain boundary conditions ensure uniqueness of solutions of the linear shallow-water equations. Verify that these same boundary conditions suffice for the nonlinear equations.   □

Exercise 6.7.2 (Bennett & Chua, 1999)

Now suppose that the domain $\mathcal{D}$ consists of the same fluid particles for all time $t > 0$, that is, the boundary $\mathcal{B}$ moves with the flow, or is "comoving". Prove that

$$\frac{dE}{dt} + C \leq BE, \qquad (6.7.11)$$

where E and B are defined as in (6.7.7) and (6.7.9), while

$$C = C(t) \equiv \int_{\mathcal{B}} gh^{(1)}h'\mathbf{u}' \cdot \hat{\mathbf{n}}\, ds. \qquad (6.7.12)$$

Prove that uniqueness follows from any one boundary condition ensuring $C \geq 0$, such as

$$\left.\begin{array}{ll}
\text{(i)} & h' = 0 \\[4pt]
\text{(ii)} & \mathbf{u}' \cdot \hat{\mathbf{n}} = 0 \\[4pt]
\text{or} & \\[4pt]
\text{(iii)} & \mathbf{u}' \cdot \hat{\mathbf{n}} = k\big(g/h^{(1)}\big)^{\frac{1}{2}} h'
\end{array}\right\}. \qquad (6.7.13)$$

The criticality of the flow at the boundary is not an issue. Indeed, the local Froude number $\mathbf{u}^{(1)} \cdot \hat{\mathbf{n}}(gh^{(1)})^{-\frac{1}{2}}$ is effectively zero in the reference frame of the comoving boundary. $\qquad\Box$

Exercise 6.7.3

How many conditions are needed in order to determine the motion of the boundary $\mathcal{B}$? $\qquad\Box$

References

Adams, R. A., 1975: *Sobolev Spaces*. Academic, New York, NY, 268 pp.

Amodei, L., 1995: Solution approchée pour un problème d'assimilation de données météorologiques avec prise en compte de l'erreur de modèle. *Comptes Rendus de l'Académie des Sciences*, **321**, Série IIa, 1087–1094.

Andersen, O. B., P. L. Woodworth, & R. A. Flather, 1995: Intercomparisons of recent ocean tide models. *J. Geophys. Res.*, **100**, 25 261–25 283.

Anderson, D. L. T., J. Sheinbaum, & K. Haines, 1996: Data assimilation in ocean models. *Rep. Progr. Phys.*, **59**, 1209–1266.

Azencott, R., 1992: *Simulated Annealing*. Wiley, New York, NY, 242 pp.

Barth, N., & C. Wunsch, 1990: Oceanographic experiment design by simulated annealing. *J. Phys. Oceanogr.*, **20**, 1249–1263.

Batchelor, G. K., 1973: *An Introduction to Fluid Dynamics*. Cambridge University Press, Cambridge, UK, 615 pp.

Bennett, A. F., 1985: Array design by inverse methods. *Progr. Oceanogr.*, **15**, 129–156.

Bennett, A. F., 1990: Inverse methods for assessing ship-of-opportunity networks and estimating circulation and winds from tropical expendable bathythermograph data. *J. Geophys. Res.*, **95**, 16 111–16 148.

Bennett, A. F., 1992: *Inverse Methods in Physical Oceanography*. Cambridge University Press, New York, NY (Reprinted 1999), 347 pp.

Bennett, A. F., & J. R. Baugh, 1992: A parallel algorithm for variational assimilation in oceanography and meteorology. *J. Atmos. Oc. Tech.*, **9**, 426–433.

Bennett, A. F., & B. S. Chua, 1994: Open-ocean modeling as an inverse problem: the primitive equations. *Mon. Wea. Rev.*, **122**, 1326–1336.

Bennett, A. F., & B. S. Chua, 1999: Open boundary conditions for Lagrangian geophysical fluid dynamics. *J. Comput. Phys.*, **153**, 418–436.

Bennett, A. F., & P. E. Kloeden, 1981: The quasigeostrophic equations: approximation, predictability and equilibrium spectra of solutions. *Q. J. R. Meteor. Soc.*, **107**, 121–136.

Bennett, A. F., & R. N. Miller, 1991: Weighting initial conditions in variational assimilation schemes. *Mon. Wea. Rev.*, **119**, 1004–1018.

Bennett, A. F., & M. A. Thorburn, 1992: The generalized inverse of a nonlinear quasigeostrophic ocean circulation model. *J. Phys. Oceanogr.*, **22**, 213–230.

Bennett, A. F., C. R. Hagelberg, & L. M. Leslie, 1992: Predicting hurricane tracks. *Nature*, **360**, 423.

Bennett, A. F., L. M. Leslie, C. R. Hagelberg, & P. E. Powers, 1993: Tropical cyclone prediction using a barotropic model initialized by a generalized inverse method. *Mon. Wea. Rev.*, **121**, 1714–1729.

Bennett, A. F., B. S. Chua, & L. M. Leslie, 1996: Generalized inversion of a global numerical weather prediction model. *Meteor. Atmos. Physics*, **60**, 165–178.

Bennett, A. F., B. S. Chua, & L. M. Leslie, 1997: Generalized inversion of a global numerical weather prediction model, II: Analysis and implementation. *Meteor. Atmos. Physics*, **62**, 129–140.

Bennett, A. F., B. S. Chua, D. E. Harrison, & M. J. McPhaden, 1998: Generalized inversion of Tropical Atmosphere–Ocean (TAO) data and a coupled model of the tropical Pacific. *J. Climate*, **11**, 1768–1792.

Bennett, A. F., B. S. Chua, D. E. Harrison, & M. J. McPhaden, 2000: Generalized inversion of Tropical Atmosphere–Ocean (TAO) data and a coupled model of the tropical Pacific, II: The 1995–96 La Niña and 1997–98 El Niño. *J. Climate*, **13**, 2770–2785.

Bleck, R., & L. T. Smith, 1990: A wind-driven isopycnic coordinate model of the North and Equatorial Atlantic Ocean. 1. Model development and supporting experiments. *J. Geophys. Res.*, **95**, 3273–3285.

Bogden, P., 2001: The impact of model–error correlations on regional data assimilative models and their observational arrays. *J. Mar. Res.*, **59**, 831–857.

Bretherton, F. P., R. E. Davis, & C. B. Fandry, 1976: A technique for the objective analysis and design of oceanographic experiments applied to MODE-73. *Deep-Sea Research*, **23**, 559–582.

Carton, J. A., G. Chepurin, & X. Cao, 2000a: A simple ocean data assimilation analysis of the global upper ocean 1950–95. Part I: Methodology. *J. Phys. Oceanogr.*, **30**, 294–309.

Carton, J. A., G. Chepurin, & X. Cao, 2000b: A simple ocean data assimilation analysis of the global upper ocean 1950–95. Part II: Results. *J. Phys. Oceanogr.*, **30**, 311–326.

Cartwright, D. E., 1999: *Tides: A Scientific History*. Cambridge University Press, Cambridge, UK, 272 pp.

Chan, N. H., J. B. Kadane, R. N. Miller, & W. Palma, 1996: Estimation of tropical sea level anomaly by an improved Kalman filter. *J. Phys. Oceanogr.*, **26**, 1286–1303.

Chandra, R., L. Dagum, D. Kohr, D. Maydan, J. McDonald, & R. Menon, 2001: *Parallel Programming in OpenMP*. Morgan Kaufmann, San Francisco, CA, 230 pp.

Chapman, S., & T. G. Cowling, 1970: *The Mathematical Theory of Non-Uniform Gases*. Cambridge University Press, Cambridge, UK, 422 pp.

Chelton, D. B., J. C. Ries, B. J. Hanes, L.-L. Fu, & P. S. Callahan, 2001: Satellite altimetry. In *Satellite Altimetry and Earth Sciences*, ed. L.-L. Fu & A. Cazanave. Academic Press, San Diego, CA, 463 pp.

Chua, B. S., & A. F. Bennett, 2001: An inverse ocean modeling system. *Ocean Modelling*, **3**, 137–165.

Cohn, S. E., 1997: An introduction to estimation theory. *J. Met. Soc. Japan*, **75**, 257–288.

Courant, R., & D. Hilbert, 1953: *Methods of Mathematical Physics*, vol. 1. Wiley Interscience, New York, NY, 560 pp.

Courant, R., & D. Hilbert, 1962: *Methods of Mathematical Physics*, vol. 2. Wiley Interscience, New York, NY, 830 pp.

Courtier, P., 1997: Dual formulation of four-dimensional assimilation. *Q. J. R. Met. Soc.*, **123**, 2449–2461.

Courtier, P., J. Derber, R. Errico, J.-F. Louis, & T. Vukicevic, 1993: Important literature on the use of adjoint, variational methods and the Kalman filter in meteorology. *Tellus*, **45A**, 342–357.

Cox, D. R., & D. V. Hinkley, 1974: *Theoretical Statistics*. Chapman and Hall, New York, NY, 511 pp.

Cox, M. D., & K. Bryan, 1984: A numerical model of the ventilated thermocline. *J. Phys. Oceanogr.*, **14**, 674–687.

Daley, R. A., 1991: *Atmospheric Data Analysis*. Cambridge University Press, Cambridge, UK, 457 pp.

Daley, R., & E. Baker, 2000: *The NAVDAS Source Book*. NRL/PU/7530-00-418. Naval Research Laboratory, Monterey, CA, 153 pp.

Davidson, N. E., & B. J. McAvaney, 1981: The ANMRC tropical analysis scheme. *Mon. Wea. Rev.*, **29**, 155–168.

Dee, D. P., & A. M. Da Silva, 1998: Data assimilation in the presence of forecast bias. *Q. J. R. Meteor. Soc.*, **124**, 269–295.

Derber, J., & A. Rosati, 1989: A global oceanic data assimilation system. *J. Phys. Oceanogr.*, **19**, 1333–1347.

Doodson, A. T., & H. D. Warburg, 1941: *Admiralty Manual of Tides*. Her Majesty's Stationery Office, London, UK, 270 pp.

Dutz, J., 1998: Repression of fecundity in the neritic copepod *Acartia clausi* exposed to the toxic dinoflagellate *Alexandrium lusitanicum*: relationship between feeding and egg production. *Mar. Ecol. Prog. Ser.*, **175**, 97–107.

Egbert, G. D., 1997: Tidal data inversion: interpolation and inference. *Progr. Oceanogr.*, **40**, 53–80.

Egbert, G. D., & A. F. Bennett, 1996: Data assimilation methods for ocean tides, in *Modern Approaches to Data Assimilation in Ocean Modeling*, ed. P. Malanotte-Rizzoli. Elsevier, New York, NY, 147–179.

Egbert, G. D., & R. D. Ray, 2000: Significant dissipation of tidal energy in the deep ocean inferred from satellite altimeter data. *Nature*, **405**, 775–778.

Egbert, G. D., & R. D. Ray, 2001: Estimates of M_2 tidal energy dissipation from TOPEX/POSEIDON altimeter data. *J. Geophys. Res.*, **106**, 22 475–22 502.

Egbert, G. D., A. F. Bennett, & M. G. G. Foreman, 1994: TOPEX/POSEIDON tides estimated using a global inverse method. *J. Geophys. Res.*, **99**, 24 821–24 852.

Eknes, M., & G. Evensen, 1997: Parameter estimation solving a weak constraint variational formulation for an Ekman model. *J. Geophys. Res.*, **102**, 12 479–12 491.

Elsberry, R. L., 1990: International experiments to study tropical cyclones in the western North Pacific. *Bull. Amer. Meteor. Soc.*, **71**, 1305–1316.

Emanuel, K., 1999: Thermodynamic control of hurricane intensity. *Nature*, **401**, 665–669.

Errico, R. M., 1997: What is an adjoint model? *Bull. Amer. Met. Soc.*, **11**, 2577–2750.

Evensen, G., 1994: Sequential data assimilation with a nonlinear quasi-geostrophic model using Monte Carlo methods to forecast error statistics. *J. Geophys. Res.*, **99**, 10 143–10 162.

Ferziger, J. H., 1996: Large eddy simulation. In *Simulation and Modeling of Turbulent Flows*, ed. T. B. Gatski, M. Y. Hussaini, & J. L. Lumley. Oxford University Press, New York, NY, 314 pp.

Fu, L. L., & I. Fukumori, 1996: A case study of the effects of errors in satellite altimetry on data assimilation, in *Modern Approaches to Data Assimilation in Ocean Modeling*, ed. P. Malanotte-Rizzoli. Elsevier, New York, NY, 77–96.

Fukumori, I., 2001: Data assimilation for models. In *Satellite Altimetry and Earth Sciences*, ed. L.-L. Fu & A. Cazanave. Academic Press, San Diego, CA, 463 pp.

Fukumori, I., & P. Malanotte-Rizzoli, 1995: An approximate Kalman filter for ocean data assimilation; an example with an idealized Gulf Stream Model. *J. Geophys. Res.*, **100**, 6777–6793.

Gelb, A. (ed.), 1974: *Applied Optimal Estimation*. MIT Press, Cambridge, MA, 374 pp.

Gelfand, I. M., & S. V. Fomin, 1963: *Calculus of Variations*. Prentice Hall, Englewood Cliffs, NJ, 232 pp.

Gent, P. R., & M. A. Cane, 1989: A reduced gravity, primitive equation model of the upper tropical ocean. *J. Comput. Phys.*, **81**, 444–480.

Ghil, M., K. Ide, A. Bennett, P. Courtier, M. Kimoto, M. Nagata, M. Saiki, & N. Sato (eds.), 1997: *Data Assimilation in Meteorology and Oceanography, Theory and Practice*. Meteorological Society of Japan, Tokyo, Japan, 496 pp.

Giering, R. & T. Kaminsky, 1997: Recipes for adjoint code construction. *ACM Trans. Math. Software*, **24**, 437–474.

Gill, A. E., 1982: *Atmosphere–Ocean Dynamics*. Academic, London, UK, 662 pp.

Golub, G. H., & C. F. Van Loan, 1989: *Matrix Computations*, 2nd ed. Johns Hopkins University Press, Baltimore, MD, 642 pp.

Goerss, J. S., & P. A. Phoebus, 1993: *The Multivariate Optimum Interpolation Analysis of Meteorological Data at the Fleet Numerical Oceanography Center*. NRL/FR/7531-92-9413. Naval Research Laboratory, Monterey, CA, 58 pp.

Greiner, E., & C. Perigaud, 1994a: Assimilation of geosat altimetric data in a nonlinear reduced-gravity model of the Indian Ocean. Part I: Adjoint approach and model-data consistency. *J. Phys. Oceanogr.*, **24**, 1783–1803.

Greiner, E., & C. Perigaud, 1994b: Assimilation of geosat altimetric data in a nonlinear reduced-gravity model of the Indian Ocean. Part II: Some validation and interpretation of the assimilated results. *J. Phys. Oceanogr.*, **26**, 1735–1746.

Hackert, E. C., R. N. Miller, & A. J. Busalacchi, 1998: An optimized design for a moored instrument in the tropical Atlantic Ocean. *J. Geophys. Res.*, **103**, 7491–7509.

Hadamard, J., 1952: *Lectures on Cauchy's Problem in Linear Partial Differential*

Equations. Dover Publications, New York, NY, 316 pp.

Haltiner, G. J., & R. T. Williams, 1980: *Numerical Prediction and Dynamical Meteorology*, 2nd ed. John Wiley, New York, NY, 477 pp.

Hoang, S., R. Baraille, O. Talagrand, X. Carton, & P. De Mey, 1997a: Adaptive filtering: application to satellite data assimilation in oceanography. *Dyn. Atmos. Oceans*, **27**, 257–281.

Hoang, S., P. De Mey, O. Talagrand, & R. Baraille, 1997b: A new reduced-order adaptive filter for state estimation. *Automatica*, **33**, 1475–1498.

Hodur, R. M., 1997: The Naval Research Laboratory's Coupled Ocean/Atmosphere Mesoscale Prediction System (COAMPS). *Mon. Wea. Rev.*, **125**, 1414–1430.

Hogan, T. F., & T. E. Rosmond, 1991: The description of the U.S. Navy Operational Global Atmospheric Prediction System's spectral forecast model. *Mon. Wea. Rev.*, **119**, 1786–1815.

Holland, G. J., 1980: An analytical model of the wind and pressure profiles in hurricanes. *Mon. Wea. Rev.*, **108**, 1212–1218.

Holton, J., 1992: *An Introduction to Dynamical Meteorology*, 3rd ed. Academic, San Diego, CA, 511 pp.

Ide, K., P. Courtier, M. Ghil, & A. Lorenc, 1997: Unified notation for data assimilation: operational, sequential and variational. *J. Met. Soc. Japan*, **75**, 181–189.

Kasibhatla, P., M. Heimann, P. Rayner, N. Mahowald, R. G. Prinn, & D. E. Hartley (eds.), 2000: *Inverse Methods in Global Biogeochemical Cycles*. Geophysical Monograph **114**, American Geophysical Union, Washington, D.C., 324 pp.

Keppenne, C. L., 2000: Data assimilation into a primitive-equation model with a parallel ensemble Kalman filter. *Mon. Wea. Rev.*, **128**, 1971–1981.

Kessler, W. S., 1990: Observations of long Rossby waves in the northern tropical Pacific. *J. Geophys. Res.*, **95**, 5183–5217.

Kessler, W. S., & J. P. McCreary, Jr., 1993: The annual wind driven Rossby waves in the subthermocline equatorial Pacific. *J. Phys. Oceanogr.*, **23**, 1192–1207.

Kessler, W. S., M. C. Spillane, M. J. McPhaden, & D. E. Harrison, 1996: Scales of variability in the equatorial Pacific inferred from the Tropical Atmosphere–Ocean buoy array. *J. Climate*, **9**, 2999–3024.

Kleeman, R. C., A. M. Moore, & N. R. Smith, 1995: Assimilation of subsurface thermal data into a simple ocean model for the initialization of an intermediate tropical coupled ocean–atmosphere forecast model. *Mon. Wea. Rev.*, **123**, 3101–3113.

Klinker, E., F. Rabier, G. Kelly, & J.-F. Mahfouf, 2000: The ECMWF operational implementation of four-dimensional variational assimilation. III: Experimental results and diagnostics with operational configuration. *Q. J. R. Meteorol. Soc.*, **126**, 1191–1215.

Kruger, J., 1993: Simulated annealing: A tool for data assimilation into an almost steady model state. *J. Phys. Oceanogr.*, **23**, 679–688.

Kundu, P. K., 1990: *Fluid Mechanics*. Academic, San Diego, CA, 638 pp.

Ladyzhenskaya, O. A., 1969: *The Mathematical Theory of Viscous Incompressible Flow*, 2nd English ed., translated from the Russian by R. A. Silverman & J. Chu. Gordon & Breach, New York, NY, 224 pp.

Lagarde, T., A. Piacentini, & O. Thual, 2001: A new representation of data-assimilation methods: The PALM flow-charting approach. *Q. J. R. Meteorol. Soc.*, **127**, 189–207.

Lanczos, C., 1966: *The Variational Principles of Mechanics*, 3rd ed. University of Toronto, Toronto, Canada, 375 pp.

Le Dimet, F., & O. Talagrand, 1986: Variational algorithms for analysis and assimilation of meteorological observations. *Tellus*, **38A**, 97–110.

Le Dimet, F.-X., H.-E. Ngodock, B. Luong, & J. Verron, 1997: Sensitivity analysis in variational data assimilation. *J. Met. Soc. Japan*, **75**, 245–255.

Le Provost, C., 2001: Ocean tides. In *Satellite Altimetry and Earth Sciences*, ed. L.-L. Fu & A. Cazanave. Academic Press, San Diego, CA, 463 pp.

Le Provost, C. L., & F. Lyard, 1997: Energetics of the barotropic ocean tides: an estimate of bottom friction dissipation from a hydrodynamic model. *Prog. Oceanogr.*, **40**, 37–52.

Le Provost, C. L., M. L. Genco, F. Lyard, P. Vincent, & P. Canceil, 1994: Spectroscopy of the world ocean tides from a hydrodynamic model. *J. Geophys. Res.*, **99**, 24 777–24 797.

Le Provost, C., A. F. Bennett, & D. E. Cartwright, 1995: Ocean tides for and from TOPEX/POSEIDON. *Science*, **267**, 639–642.

Lewis, J. M., & J. C. Derber, 1985: The use of adjoint equations to solve a variational adjustment problem with advective constraints. *Tellus*, **37A**, 309–322.

Lions, J. L., 1971: *Optimal Control of Systems Governed by Partial Differential Equations*. Springer-Verlag, Berlin, Federal Republic of Germany, 396 pp.

Ljung, L., & T. Söderström, 1987: *Theory and Practice of Recursive Identification*. MIT Press, Cambridge, MA, 529 pp.

Lorenc, A., 1997: Atmospheric data assimilation. Forecasting Research Scientific Paper 34, Meteorological Office, Bracknell, UK.

Lorenc, A. C., S. P. Ballard, R. S. Bell, N. B. Ingleby, P. L. F. Andrews, D. M. Barker, J. R. Bray, A. M. Clayton, T. Dalby, D. Li, T. J. Payne, & F. W. Saunders, 2000: The Met. Office global three-dimensional variational data assimilation scheme. *Q. J. R. Meteorol. Soc.*, **126**, 2991–3012.

Lyard, F. H., 1999: Data assimilation in a wave equation: a variational approach for the Grenoble tidal model. *J. Comp. Phys.*, **149**, 1–31.

Lyard, F. H., & C. Le Provost, 1997: Energy budget of the tidal hydrodynamic model FES 94.1. *Geophys. Res. Lett.*, **24**, 687–690.

Mahfouf, J.-F., & F. Rabier, 2000: The ECMWF operational implementation of four-dimensional variational assimilation. II: Experimental results with improved physics. *Q. J. R. Meteorol. Soc.*, **126**, 1171–1190.

Malanotte-Rizzoli, P. (ed.), 1996: *Modern Approaches to Data Assimilation in Ocean Modeling*. Elsevier, Amsterdam, The Netherlands, 445 pp.

Malanotte-Rizzoli, P., I. Fukumori, & R. E. Young, 1996: A methodology for the construction of a hierarchy of Kalman filters for nonlinear primitive equation models. In *Modern Approaches to Data Assimilation in Ocean Modeling*, ed. P. Malanotte-Rizzoli. Elsevier, New York, NY, 297–317.

Marotske, J., R. Giering, K. Q. Zhang, D. Stammer, C. Hill, & T. Lee, 1999: Construction of the adjoint MIT ocean general circulation model and application to Atlantic heat transport sensitivity. *J. Geophys. Res.*, **104**, 29 529–29 547.

Marshall, J., C. Hill, L. Perelman, & A. Adcroft, 1997a: Hydrostatic, quasi-hydrostatic, and non-hydrostatic ocean modeling. *J. Geophys. Res.*, **102**, 5733–5752.

Marshall, J., A. Adcroft, C. Hill, L. Perelman, & C. Heisey, 1997b: A finite-volume, incompressible Navier–Stokes model for studies of the ocean on parallel computers. *J. Geophys. Res.*, **102**, 5753–5766.

McIntosh, P. C., & A. F. Bennett, 1984: Open ocean modeling as an inverse problem: M_2 tides in Bass Strait. *J. Phys. Oceanogr.*, **14**, 601–614.

McPhaden, M. J., 1993: TOGA–TAO and the 1991–93 El Niño Southern Oscillation event. *Oceanography*, **6**, 36–44.

McPhaden, M. J., 1999a: Genesis and evolution of the 1997–98 El Niño. *Science*, **283**, 950–954.

McPhaden, M. J., 1999b: El Niño: the child prodigy of 1997–98. *Nature*, **398**, 559–562.

Meditch, J. S., 1970: On state estimation for distributed parameter systems. *J. Franklin Inst.*, **290**, 45–59.

Mesinger, F., & A. Arakawa, 1976: *Numerical Methods Used in Atmospheric Models.* GARP Publication Series No. 14, WMO/ICSU Joint Organizing Committee, Geneva, Switzerland, 64 pp.

Metropolis, N., A. W. Rosenbluth, M. N. Rosenbluth, A. H. Teller, & E. Teller, 1953: Equation of state calculation by fast computing machines. *J. Chem. Physics*, **21**, 1087–1092.

Miller, R. N., 1996: *Introduction to the Kalman Filter.* ECMWF Workshop on Data Assimilation, Reading, UK.

Miller, R. N., E. D. Zaron, & A. F. Bennett, 1994: Data assimilation in models with convective adjustment. *Mon. Wea. Rev.*, **122**, 2607–2613.

Miller, R. N., A. J. Busalacchi, & E. C. Hackert, 1995: Sea surface topography fields of the tropical Pacific from data assimilation. *J. Geophys. Res.*, **100**, 13 389–13 425.

Muccino, J. C., & A. F. Bennett, 2001: Generalized inversion of the Korteweg–deVries equation. *Dyn. Atmos. Oceans*, (to appear).

Oliger, J., & A. Sundström, 1978: Theoretical and practical aspects of some initial boundary value problems in fluid dynamics, *SIAM J. App. Math.*, **35**, 419–446.

Pacanowski, R. C., & S. G. H. Philander, 1981: Parameterization of vertical mixing in numerical models of tropical oceans. *J. Phys. Oceanogr.*, **11**, 1443–1451.

Pacheco, P., 1996: *Parallel Programming with MPI.* Morgan Kaufmann, San Francisco, CA, 440 pp.

Parker, R. L., 1994: *Geophysical Inverse Theory.* Princeton University Press, Princeton, NJ, 386 pp.

Passi, R. M., R. K. Goodrich, & J. C. Derber, 1996: A convolution algorithm with application to data assimilation. *SIAM J. Sci. Comput.*, **17**, 942–955.

Pedlosky, J., 1987: *Geophysical Fluid Dynamics*, 2nd ed. Springer-Verlag, New York, NY, 710 pp.

Press, W. H., S. A. Teukolsky, W. T. Vetterling, & B. P. Flannery, 1986: *Numerical Recipes, the Art of Scientific Computing.* Cambridge University Press, New York, NY, 735 pp.

Rabier, F., H. Järvinen, E. Klinker, J.-F. Mahfouf, & A. Simmons, 2000: The ECMWF operational implementation of four-dimensional assimilation. I: Experimental results with simplified physics. *Q. J. R. Meteorol. Soc.*, **126**, 1143–1170.

Reichle, R. H., D. Entekhabi, & D. B. McLaughlin, 2001: Downscaling of radiobrightness measurements for soil moisture estimation: a four-dimensional variational data assimilation approach. *Water Resource Res.*, **37**, 2353–2364.

Reichle, R. H., D. B. McLaughlin, & D. Entekhabi, 2002: Hydrologic data assimilation with the Ensemble Kalman filter. *Mon. Wea. Rev.*, **130**, 103–114.

Reid, W. T., 1968: Generalized inverses of differential and integral operators. In *Theory and Applications of Generalized Inverses of Matrices*, ed. T. L. Boullion & P. L. Odell, pp. 1–25. Texas Technical College, Lubbock, TX.

Reynolds, R. W., & T. M. Smith, 1995: A high resolution global sea surface temperature climatology. *J. Climate*, **8**, 1571–1583.

Rodgers, C. D., 2000: *Inverse Methods for Atmospheric Sounding: Theory and Practice*. World Scientific, Singapore, 256 pp.

Sasaki, Y., 1970: Some basic formalisms in numerical variational analysis. *Mon. Wea. Rev.*, **98**, 875–883.

Schiff, L. I., 1949: *Quantum Mechanics*. McGraw-Hill, New York, NY, 404 pp.

Schrama, E. J. O., & R. Ray, 1994: A preliminary tidal analysis of TOPEX/POSEIDON altimetry. *J. Geophys. Res.*, **99**, 24 799–24 808.

Shum, C. K., P. L. Woodworth, O. B. Andersen, G. Egbert, O. Francis, C.

King, S. Klosko, C. Le Provost, X. Li, J. M. Molines, M. Parke, & C. Wunsch, 1997: Accuracy assessment of recent ocean tidal models. *J. Geophys. Res.*, **102**, 25 173–25 194.

Soreide, N. N., D. C. McClurg, W. H. Zhu, M. J. McPhaden, D. W. Denbo, & M. W. Renton, 1996: World Wide Web access to real-time and historical data from the TAO array of moored buoys in the tropical Pacific Ocean: updates for 1996. *Proc. OCEANS 96*, Fort Lauderdale, FL, MTS/IEEE, 1354–1359.

Stakgold, I., 1979: *Green's Functions and Boundary Value Problems*. Wiley, New York, NY, 638 pp.

Stammer, D., R. Davis, L.-L. Fu, I. Fukumori, R. Giering, T. Lee, J. Marotzke, J. Marshall, D. Menemenlis, P. Niiler, C. Wunsch, & V. Zlotnicki, 2000: Ocean state estimation in support of CLIVAR and GODAE. *CLIVAR Exchanges*, **5**, 3–5. www.clivar.org

Thacker, W. C., & R. B. Long, 1988: Fitting dynamics to data. *J. Geophys. Res.*, **93**, 1227–1240.

Thiébaux, H. J., & M. A. Pedder, 1987: *Spatial Objective Analysis*. Academic, New York, NY, 299 pp.

Uboldi, F., & M. Kamachi, 2000: Time–space weak-constraint data assimilation for nonlinear models. *Tellus*, **52A**, 412–421.

Velden, C. S., C. M. Hayden, W. P. Menzel, J. L. Franklin, & J. S. Lynch, 1992: The impact of satellite-derived winds on numerical hurricane track forecasts. *Wea. & Forecasting*, **7**, 107–118.

Wahba, G., & J. Wendelberger, 1980: Some new mathematical methods for variational objective analysis using splines and cross-validation. *Mon. Wea. Rev.*, **108**, 1122–1143.

Wallace, J. M., & P. V. Hobbs, 1977: *Atmospheric Science: An Introductory Survey*. Academic, New York, NY, 467 pp.

Weinert, H. L. (ed.), 1982: *Reproducing Kernel Hilbert Spaces, Applications in Statistical Signal Processing*. Hutchinson Ross, Stroudsburg, PA, 654 pp.

Woodruff, S. D., R. J. Slutz, R. L. Jenne, & P. Steurer, 1987: A comprehensive atmosphere–ocean data set. *Bull. Amer. Meteor. Soc.*, **68**, 1239–1250.

Wunsch, C., 1996: *The Ocean Circulation Inverse Problem*. Cambridge University Press, New York, NY, 442 pp.

Wunsch, C., 1998: The work done by the wind on the ocean circulation. *J. Phys. Oceanogr.*, **28**, 2331–2339.

Xu, L., & R. Daley, 2000: Towards a true 4-dimensional data assimilation algorithm: application of a cycling representer algorithm to a simple transport problem. *Tellus*, **52A**, 109–128.

Xu, Q., & J. Gao, 1999: Generalized adjoint for physical processes with parameterized discontinuities. Part IV: Minimization problems in multidimensional space. *J. Atmos. Sci.*, **56**, 994–1002.

Yoshida, K., 1980: *Functional Analysis*, 6th ed. Springer-Verlag, Berlin, Federal Republic of Germany, 496 pp.

Zebiak, S. E., & M. A. Cane, 1987: A model El Niño/Southern Oscillation. *Mon. Wea. Rev.*, **115**, 2262–2278.

Zhang, S., X. Zou, J. Ahlquist, & I. M. Navon, 2000: Use of differentiable and nondifferentiable optimization algorithms for variational data assimilation with discontinuous cost functions. *Mon. Wea. Rev.*, **128**, 4031–4044.

Appendix A

Computing exercises

These computational exercises complement the analytical development of variational data assimilation in the text, and also serve to develop the confidence needed for more ambitious calculations. All code for these exercises is available at an anonymous ftp site. The linear, one-dimensional "toy" model of §1.1 is upgraded here to a linear, two-dimensional shallow-water model in both continuous and finite-difference form. Continuous and discrete penalty functionals are developed, and the respective Euler–Langrange equations are derived. Representers are calculated directly, so the generalized inverse may then be calculated directly or indirectly.

A.1 Forward model

A.1.1 Preamble

The exercise is to construct and run a simple forward model using standard numerical methods. You may obtain the source code from our anonymous `ftp` site (`ftp ftp.oce.orst.edu`, then `cd /dist/bennett/class`), along with some plotting utilities. It is assumed that your computer environment is UNIX, with the Fortran compilation command 'f77'.

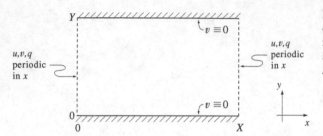

Figure A.1.1 Periodic channel with rigid walls, rotating at the rate $\frac{f}{2}$ about an axis normal to the xy-plane.

A.1.2   Model

We consider a linear shallow-water model

$$\frac{\partial u}{\partial t} - fv + g\frac{\partial q}{\partial x} + r_u u = F_u,$$

$$\frac{\partial v}{\partial t} + fu + g\frac{\partial q}{\partial y} + r_v v = F_v,$$

$$\frac{\partial q}{\partial t} + H\left(\frac{\partial u}{\partial x} + \frac{\partial v}{\partial y}\right) + r_q q = 0,$$

on the domain $0 \leq x \leq X$ and $0 \leq y \leq Y$ (see Fig. A.1.1).

A.1.3   Initial conditions

The initial values are

$$u(x, y, 0) = I^u(x, y) = 0,$$
$$v(x, y, 0) = I^v(x, y) = 0,$$

and

$$q(x, y, 0) = I^q(x, y) = 0.$$

A.1.4   Boundary conditions

The north and south walls are rigid:

$$v(x, 0, t) = v(x, Y, t) = 0,$$

while all fields are periodic in the x-direction:

$$u(x \pm X, y, t) = u(x, y, t),$$
$$v(x \pm X, y, t) = v(x, y, t),$$

and

$$q(x \pm X, y, t) = q(x, y, t).$$

A.1.5   Model forcings

The model forcings are

$$F_u = -C_d\rho_a u_a^2/(H\rho_w),$$

and

$$F_v = 0.$$

A.1.6   Model parameters

The following parameters are suggested:

zonal period, $X = 2000$ km
meridional width, $Y = 1000$ km
mean depth, $H = 5000$ m
time interval, $T = 1.8 \times 10^4$ s
gravitational acceleration, $g = 9.806$ m s^{-2}
Coriolis parameter, $f = 1.0 \times 10^{-4}$ s^{-1}
damping coefficient, $r_u = (1.8 \times 10^4$ s$)^{-1}$
damping coefficient, $r_v = (1.8 \times 10^4$ s$)^{-1}$
damping coefficient, $r_q = (1.8 \times 10^4$ s$)^{-1}$
drag coefficient, $C_d = 1.6 \times 10^{-3}$
air density, $\rho_a = 1.275$ kg m^{-3}
water density, $\rho_w = 1.0 \times 10^3$ kg m^{-3}
zonal wind, $u_a = 5$ m s^{-1}.

A.1.7   Numerical model

The differential equations are discretized on the Arakawa C-grid (see Fig. A.1.2) with
a forward–backward scheme for time-stepping (Mesinger & Arakawa, 1976) given as
follows:

$$\frac{q_{i,j}^{k+1} - q_{i,j}^k}{\Delta t} + H\left(\frac{u_{i+1,j}^k - u_{i,j}^k}{\Delta x} + \frac{v_{i,j+1}^k - v_{i,j}^k}{\Delta y}\right) + r_q q_{i,j}^k = 0,$$

$$\frac{u_{i,j}^{k+1} - u_{i,j}^k}{\Delta t} - f\left(\frac{v_{i,j+1}^k + v_{i,i}^k + v_{i-1,j+1}^k + v_{i-1,j}^k}{4}\right)$$

$$+ g\left(\frac{q_{i,j}^{k+1} - q_{i-1,j}^{k+1}}{\Delta x}\right) + r_u u_{i,j}^k = F_{u_{i,j}^k},$$

$$\frac{v_{i,j}^{k+1} - v_{i,j}^k}{\Delta t} + f\left(\frac{u_{i+1,j}^k + u_{i,j}^k + u_{i+1,j-1}^k + u_{i,j-1}^k}{4}\right)$$

$$+ g\left(\frac{q_{i,j}^{k+1} - q_{i,j-1}^{k+1}}{\Delta y}\right) + r_v v_{i,j}^k = F_{v_{i,j}^k},$$

Spatial:

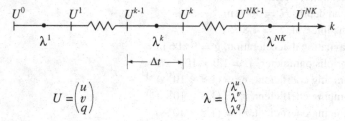

Temporal:

$$U = \begin{pmatrix} u \\ v \\ q \end{pmatrix} \qquad \lambda = \begin{pmatrix} \lambda^u \\ \lambda^v \\ \lambda^q \end{pmatrix}$$

Figure A.1.2 Arakawa C-grid for space differences; staggering of forward and adjoint variables for time differences.

where

$$q_{i,j}^k : i = 1, NI; \quad j = 1, NJ - 1; \quad k = 0, NK - 1,$$
$$u_{i,j}^k : i = 1, NI; \quad j = 1, NJ - 1; \quad k = 0, NK - 1,$$

and

$$v_{i,j}^k : i = 1, NI; \quad j = 2, NJ - 1; \quad k = 0, NK - 1.$$

Rigid boundary conditions

$$v_{i,1}^k = 0,$$

and

$$v_{i,NJ}^k = 0.$$

Periodic boundary conditions

$$u_{0,j}^k = u_{NI,j}^k,$$
$$u_{NI+1,j}^k = u_{1,j}^k,$$
$$v_{0,j}^k = v_{NI,j}^k,$$
$$v_{NI+1,j}^k = v_{1,j}^k,$$
$$q_{0,j}^k = q_{NI,j}^k,$$

and

$$q_{NI+1,j}^k = q_{1,j}^k.$$

A.1.8 Numerical model parameters

The following parameters are suggested:

number of grid points in x-direction, $NI = 20$
number of grid points in y-direction, $NJ = 11$
number of time steps, $NK = 100$
grid spacing (x-direction), $\Delta x = 100$ km
grid spacing (y-direction), $\Delta y = 100$ km
time step, $\Delta t = 180$ s.

A.1.9 Numerical code and output

Source code: fwd.f

To compile the source code: `f77-O3 fwd.f -o fwd`
Output file: ufwd.dat, vfwd.dat and qfwd.dat

A.1.10 To generate a plot

(i) *Postprocessing*
Source code: postprocess.f
To compile the source code: `f77 postprocess.f -o postprocess`
Input file: ufwd.dat, vfwd.dat and qfwd.dat
Output file: u.dat, v.dat, q.dat, uy.dat, vy.dat, qy.dat
(ii) *Line plot*
`gnuplot line.gnu` (input file: qy.dat; output file: qy.ps)
(iii) *Contour plot*
`gnuplot contour.gnu` (input file: q.dat; output file: q.ps)

A.2 Variational data assimilation

A.2.1 Preamble

The exercise is to reformulate the forward model of Exercise A.1 as an inverse model, to define a penalty functional, to derive the associated system of Euler–Lagrange (EL) equations and to express its solution using representers.

A.2.2 Penalty functional

Formulate the penalty functional $\mathcal{J}$ for the following problem (see §1.2, Inverse models):

Equations of motion

$$\frac{\partial u}{\partial t} - fv + g\frac{\partial q}{\partial x} + r_u u = F_u + f^u,$$

$$\frac{\partial v}{\partial t} + fu + g\frac{\partial q}{\partial y} + r_v v = F_v + f^v,$$

$$\frac{\partial q}{\partial t} + H\left(\frac{\partial u}{\partial x} + \frac{\partial v}{\partial y}\right) + r_q q = f^q.$$

Initial conditions

$$u(x, y, 0) = I^u(x, y) + i^u(x, y),$$
$$v(x, y, 0) = I^v(x, y) + i^v(x, y),$$
$$q(x, y, 0) = I^q(x, y) + i^q(x, y).$$

Rigid boundary conditions

$$v(x, 0, t) = b_o(x, t),$$
$$v(x, Y, t) = b_Y(x, t).$$

Periodic boundary conditions

$$u(x \pm X, y, t) = u(x, y, t),$$
$$v(x \pm X, y, t) = v(x, y, t),$$
$$q(x \pm X, y, t) = q(x, y, t).$$

Data

$$d_m = q(x_m, y_m, t_m) + \epsilon_m, \quad 1 \le m \le M.$$

A.2.3 Euler–Lagrange equations

Derive EL equations for the extremum of the penalty functional $\mathcal{J}$.

A.2.4 Representer solution

Solve the EL equations using representers (see §1.3: Solving the Euler–Lagrange equations using representers).

A.2.5 Solutions for A.2.2–A.2.4

Penalty functional, in terms of residuals:

$$
\begin{aligned}
\mathcal{J} &= \mathcal{J}[u, v, q] \\
&= W_f^u \int_0^T dt \int_0^X dx \int_0^Y dy (f^u(x, y, t))^2 + W_f^v \int_0^T dt \int_0^X dx \int_0^Y dy (f^v(x, y, t))^2 \\
&\quad + W_f^q \int_0^T dt \int_0^X dx \int_0^Y dy (f^q(x, y, t))^2 + W_i^u \int_0^X dx \int_0^Y dy (i^u(x, y))^2 \\
&\quad + W_i^v \int_0^X dx \int_0^Y dy (i^v(x, y))^2 + W_i^q \int_0^X dx \int_0^Y dy (i^q(x, y))^2 \\
&\quad + W_b^v \int_0^T dt \int_0^X dx (b_\circ(x, t))^2 + W_b^v \int_0^T dt \int_0^X dx (b_r(x, t))^2 + w \sum_{m=1}^M (\epsilon_m)^2.
\end{aligned}
$$

Penalty functional, in terms of state variables:

$$
\begin{aligned}
\mathcal{J} &= \mathcal{J}[u, v, q] \\
&= W_f^u \int_0^T dt \int_0^X dx \int_0^Y dy \left\{ \frac{\partial u}{\partial t} - fv + g\frac{\partial q}{\partial x} + r_u u - F_u \right\}^2 \\
&\quad + W_f^v \int_0^T dt \int_0^X dx \int_0^Y dy \left\{ \frac{\partial v}{\partial t} + fu + g\frac{\partial q}{\partial y} + r_v v - F_v \right\}^2 \\
&\quad + W_f^q \int_0^T dt \int_0^X dx \int_0^Y dy \left\{ \frac{\partial q}{\partial t} + H\left(\frac{\partial u}{\partial x} + \frac{\partial v}{\partial y} \right) + r_q q \right\}^2 \\
&\quad + W_i^u \int_0^X dx \int_0^Y dy \{u(x, y, 0) - I^u(x, y)\}^2 \\
&\quad + W_i^v \int_0^X dx \int_0^Y dy \{v(x, y, 0) - I^v(x, y)\}^2 \\
&\quad + W_i^q \int_0^X dx \int_0^Y dy \{q(x, y, 0) - I^q(x, y)\}^2
\end{aligned}
$$

$$+ W_b^v \int\limits_0^T dt \int\limits_0^X dx \{v(x, 0, t)\}^2$$

$$+ W_b^v \int\limits_0^T dt \int\limits_0^X dx \{v(x, Y, t)\}^2$$

$$+ w \sum_{m=1}^M \{q(x_m, y_m, t_m) - d_m\}^2.$$

Weighted residuals

$$\lambda^u \equiv W_f^u \left(\frac{\partial \hat{u}}{\partial t} - f\hat{v} + g\frac{\partial \hat{q}}{\partial x} + r_u \hat{u} - F_u \right),$$

$$\lambda^v \equiv W_f^v \left(\frac{\partial \hat{v}}{\partial t} + f\hat{u} + g\frac{\partial \hat{q}}{\partial y} + r_v \hat{v} - F_v \right),$$

$$\lambda^q \equiv W_f^q \left(\frac{\partial \hat{q}}{\partial t} + H \left(\frac{\partial \hat{u}}{\partial x} + \frac{\partial \hat{v}}{\partial y} \right) + r_q \hat{q} \right).$$

Euler–Lagrange equations

$$-\frac{\partial \lambda^u}{\partial t} + f\lambda^v - H\frac{\partial \lambda^q}{\partial x} + r_u \lambda^u = 0,$$

$$-\frac{\partial \lambda^v}{\partial t} - f\lambda^u - H\frac{\partial \lambda^q}{\partial y} + r_v \lambda^v = 0,$$

$$-\frac{\partial \lambda^q}{\partial t} - g\left(\frac{\partial \lambda^u}{\partial x} + \frac{\partial \lambda^v}{\partial y} \right) + r_q \lambda^q$$

$$= -w \sum_{m=1}^M (\hat{q}(x_m, y_m, t_m) - d_m)\, \delta(x - x_m)\delta(y - y_m)\delta(t - t_m).$$

$$\lambda^u(x, y, T) = 0,$$
$$\lambda^v(x, y, T) = 0,$$
$$\lambda^q(x, y, T) = 0.$$

$$\lambda^v(x, 0, t) = 0,$$
$$\lambda^v(x, Y, t) = 0.$$

$$\lambda^u(x \pm X, y, t) = \lambda^u(x, y, t),$$
$$\lambda^v(x \pm X, y, t) = \lambda^v(x, y, t),$$
$$\lambda^q(x \pm X, y, t) = \lambda^q(x, y, t).$$

$$\frac{\partial \hat{u}}{\partial t} - f\hat{v} + g\frac{\partial \hat{q}}{\partial x} + r_u \hat{u} = F_u + \left[W_f^u\right]^{-1} \lambda^u,$$

$$\frac{\partial \hat{v}}{\partial t} + f\hat{u} + g\frac{\partial \hat{q}}{\partial y} + r_v \hat{v} = F_v + \left[W_f^v\right]^{-1} \lambda^v,$$

$$\frac{\partial \hat{q}}{\partial t} + H\left(\frac{\partial \hat{u}}{\partial x} + \frac{\partial \hat{v}}{\partial y}\right) + r_q \hat{q} = \left[W_f^q\right]^{-1} \lambda^q.$$

$$\hat{u}(x, y, 0) = I^u(x, y) + \left[W_i^u\right]^{-1} \lambda^u(x, y, 0),$$

$$\hat{v}(x, y, 0) = I^v(x, y) + \left[W_i^v\right]^{-1} \lambda^v(x, y, 0),$$

$$\hat{q}(x, y, 0) = I^q(x, y) + \left[W_i^q\right]^{-1} \lambda^q(x, y, 0).$$

$$\hat{v}(x, 0, t) = H\left[W_b^v\right]^{-1} \lambda^q(x, 0, t),$$

$$\hat{v}(x, Y, t) = -H\left[W_b^v\right]^{-1} \lambda^q(x, Y, t).$$

$$\hat{u}(x \pm X, y, t) = \hat{u}(x, y, t),$$

$$\hat{v}(x \pm X, y, t) = \hat{v}(x, y, t),$$

$$\hat{q}(x \pm X, y, t) = \hat{q}(x, y, t).$$

First guess

$$\frac{\partial u_F}{\partial t} - f v_F + g\frac{\partial q_F}{\partial x} + r_u u_F = F_u,$$

$$\frac{\partial v_F}{\partial t} + f u_F + g\frac{\partial q_F}{\partial y} + r_v v_F = F_v,$$

$$\frac{\partial q_F}{\partial t} + H\left(\frac{\partial u_F}{\partial x} + \frac{\partial v_F}{\partial y}\right) + r_q q_F = 0.$$

$$u_F(x, y, 0) = I^u(x, y),$$

$$v_F(x, y, 0) = I^v(x, y),$$

$$q_F(x, y, 0) = I^q(x, y).$$

$$v_F(x, 0, t) = 0,$$

$$v_F(x, Y, t) = 0.$$

$$u_F(x \pm X, y, t) = u_F(x, y, t),$$

$$v_F(x \pm X, y, t) = v_F(x, y, t),$$

$$q_F(x \pm X, y, t) = q_F(x, y, t).$$

Representer adjoint equations

$$-\frac{\partial \alpha_m^u}{\partial t} + f\alpha_m^v - H\frac{\partial \alpha_m^q}{\partial x} + r_u\alpha_m^u = 0,$$

$$-\frac{\partial \alpha_m^v}{\partial t} - f\alpha_m^u - H\frac{\partial \alpha_m^q}{\partial y} + r_v\alpha_m^v = 0,$$

$$-\frac{\partial \alpha_m^q}{\partial t} - g\left(\frac{\partial \alpha_m^u}{\partial x} + \frac{\partial \alpha_m^v}{\partial y}\right) + r_q\alpha_m^q = \delta(x - x_m)\delta(y - y_m)\delta(t - t_m).$$

$$\alpha_m^u(x, y, T) = 0,$$
$$\alpha_m^v(x, y, T) = 0,$$
$$\alpha_m^q(x, y, T) = 0.$$

$$\alpha_m^v(x, 0, t) = 0,$$
$$\alpha_m^v(x, Y, t) = 0.$$

$$\alpha_m^u(x \pm X, y, t) = \alpha_m^u(x, y, t),$$
$$\alpha_m^v(x \pm X, y, t) = \alpha_m^v(x, y, t),$$
$$\alpha_m^q(x \pm X, y, t) = \alpha_m^q(x, y, t).$$

Representer equations

$$\frac{\partial r_m^u}{\partial t} - fr_m^v + g\frac{\partial r_m^q}{\partial x} + r_u r_m^u = \left[W_f^u\right]^{-1}\alpha_m^u,$$

$$\frac{\partial r_m^v}{\partial t} + fr_m^u + g\frac{\partial r_m^q}{\partial y} + r_v r_m^v = \left[W_f^v\right]^{-1}\alpha_m^v,$$

$$\frac{\partial r_m^q}{\partial t} + H\left(\frac{\partial r_m^u}{\partial x} + \frac{\partial r_m^v}{\partial y}\right) + r_q r_m^q = \left[W_f^q\right]^{-1}\alpha_m^q.$$

$$r_m^u(x, y, 0) = \left[W_i^u\right]^{-1}\alpha_m^u(x, y, 0),$$

$$r_m^v(x, y, 0) = \left[W_i^v\right]^{-1}\alpha_m^v(x, y, 0),$$

$$r_m^q(x, y, 0) = \left[W_i^q\right]^{-1}\alpha_m^q(x, y, 0).$$

$$r_m^v(x, 0, t) = H\left[W_b^v\right]^{-1}\alpha_m^q(x, 0, t),$$

$$r_m^v(x, Y, t) = -H\left[W_b^v\right]^{-1}\alpha_m^q(x, Y, t).$$

$$r_m^u(x \pm X, y, t) = r_m^u(x, y, t),$$
$$r_m^v(x \pm X, y, t) = r_m^v(x, y, t),$$
$$r_m^q(x \pm X, y, t) = r_m^q(x, y, t).$$

Extremum of $\mathcal{J}$

$$\hat{u}(x, y, t) = u_F(x, y, t) + \sum_{m=1}^{M} \hat{\beta}_m r_m^u(x, y, t),$$

$$\hat{v}(x, y, t) = v_F(x, y, t) + \sum_{m=1}^{M} \hat{\beta}_m r_m^v(x, y, t),$$

$$\hat{q}(x, y, t) = q_F(x, y, t) + \sum_{m=1}^{M} \hat{\beta}_m r_m^q(x, y, t),$$

$$\sum_{l=1}^{M} \left(r_{lm}^q + w^{-1}\delta_{lm}\right) \hat{\beta}_l = h_m \quad (m = 1, M),$$

where

$$r_{lm} = r_l(x_m, y_m, t_m),$$

and

$$h_m = d_m - q_F(x_m, y_m, t_m).$$

A.3 Discrete formulation

A.3.1 Preamble

Verify the discrete formulation given here in detail. Derive the corresponding equations for the representers and their adjoints. Compare with the source code rep.f.

A.3.2 Penalty functional

$$\mathcal{J} = \mathcal{J}[u, v, q] = W_f^u \sum_{k=0}^{NK-1} \sum_{j=1}^{NJ-1} \sum_{i=1}^{NI} \left((f^u)_{i,j}^{k+1}\right)^2 \Delta x \Delta y \Delta t$$

$$+ W_f^v \sum_{k=0}^{NK-1} \sum_{j=2}^{NJ-1} \sum_{i=1}^{NI} \left((f^v)_{i,j}^{k+1}\right)^2 \Delta x \Delta y \Delta t$$

$$+ W_f^q \sum_{k=0}^{NK-1} \sum_{j=1}^{NJ-1} \sum_{i=1}^{NI} \left((f^q)_{i,j}^{k+1}\right)^2 \Delta x \Delta y \Delta t$$

$$+ W_i^u \sum_{j=1}^{NJ-1} \sum_{i=1}^{NI} ((i^u)_{i,j})^2 \Delta x \Delta y$$

$$+ W_i^v \sum_{j=1}^{NJ} \sum_{i=1}^{NI} ((i^v)_{i,j})^2 \Delta x \Delta y$$

$$+ W_i^q \sum_{j=1}^{NJ-1} \sum_{i=1}^{NI} ((i^q)_{i,j})^2 \Delta x \Delta y$$

$$+ W_b^v \sum_{k=1}^{NK} \sum_{i=1}^{NI} ((b_0)_i^k)^2 \Delta x \Delta t$$

$$+ W_b^v \sum_{k=1}^{NK} \sum_{i=1}^{NI} ((b_Y)_i^k)^2 \Delta x \Delta t$$

$$+ w \sum_{m=1}^{M} (\epsilon_m)^2,$$

where

$$(f^u)_{i,j}^{k+1} = \frac{u_{i,j}^{k+1} - u_{i,j}^k}{\Delta t} - f \left(\frac{v_{i,j+1}^k + v_{i,j}^k + v_{i-1,j+1}^k + v_{i-1,j}^k}{4} \right)$$

$$+ g \left(\frac{q_{i,j}^{k+1} - q_{i-1,j}^{k+1}}{\Delta x} \right) + r_u u_{i,j}^k - (F^u)_{i,j}^k,$$

$$(f^v)_{i,j}^{k+1} = \frac{v_{i,j}^{k+1} - v_{i,j}^k}{\Delta t} + f \left(\frac{u_{i+1,j}^k + u_{i,j}^k + u_{i+1,j-1}^k + u_{i,j-1}^k}{4} \right)$$

$$+ g \left(\frac{q_{i,j}^{k+1} - q_{i,j-1}^{k+1}}{\Delta y} \right) + r_v v_{i,j}^k - (F^v)_{i,j}^k,$$

$$(f^q)_{i,j}^{k+1} = \frac{q_{i,j}^{k+1} - q_{i,j}^k}{\Delta t} + H \left(\frac{u_{i+1,j}^k - u_{i,j}^k}{\Delta x} + \frac{v_{i,j+1}^k - v_{i,j}^k}{\Delta y} \right) + r_q q_{i,j}^k,$$

$$(i^u)_{i,j} = u_{i,j}^0 - (I^u)_{i,j},$$

$$(i^v)_{i,j} = v_{i,j}^0 - (I^v)_{i,j},$$

$$(i^q)_{i,j} = q_{i,j}^0 - (I^q)_{i,j},$$

$$(b_0)_i^k = v_{i,1}^k,$$

$$(b_Y)_i^k = v_{i,NJ}^k,$$

$$\epsilon_m = q_{i_m,j_m}^{k_m} - d_m.$$

A.3.3   Weighted residuals

$$(\lambda^u)_{i,j}^{k+1} \equiv W_f^u \left\{ \frac{\hat{u}_{i,j}^{k+1} - \hat{u}_{i,j}^k}{\Delta t} - f \left(\frac{\hat{v}_{i,j+1}^k + \hat{v}_{i,j}^k + \hat{v}_{i-1,j+1}^k + \hat{v}_{i-1,j}^k}{4} \right) \right.$$

$$\left. + g \left(\frac{\hat{q}_{i,j}^{k+1} - \hat{q}_{i-1,j}^{k+1}}{\Delta x} \right) + r_u \hat{u}_{i,j}^k - (F^u)_{i,j}^k \right\},$$

$$(\lambda^v)_{i,j}^{k+1} \equiv W_f^v \left\{ \frac{\hat{v}_{i,j}^{k+1} - \hat{v}_{i,j}^k}{\Delta t} + f \left(\frac{\hat{u}_{i+1,j}^k + \hat{u}_{i,j}^k + \hat{u}_{i+1,j-1}^k + \hat{u}_{i,j-1}^k}{4} \right) \right.$$

$$\left. + g \left(\frac{\hat{q}_{i,j}^{k+1} - \hat{q}_{i,j-1}^{k+1}}{\Delta y} \right) + r_v \hat{v}_{i,j}^k - (F^v)_{i,j}^k \right\},$$

$$(\lambda^q)_{i,j}^{k+1} \equiv W_f^q \left\{ \frac{\hat{q}_{i,j}^{k+1} - \hat{q}_{i,j}^k}{\Delta t} + H \left(\frac{\hat{u}_{i+1,j}^k - \hat{u}_{i,j}^k}{\Delta x} + \frac{\hat{v}_{i,j+1}^k - \hat{v}_{i,j}^k}{\Delta y} \right) + r_q \hat{q}_{i,j}^k \right\}.$$

Euler–Lagrange equations

$$-\frac{(\lambda^u)_{i,j}^{k+1} - (\lambda^u)_{i,j}^k}{\Delta t} + f \left(\frac{(\lambda^v)_{i,j+1}^{k+1} + (\lambda^v)_{i,j}^{k+1} + (\lambda^v)_{i-1,j+1}^{k+1} + (\lambda^v)_{i-1,j}^{k+1}}{4} \right)$$

$$- H \left(\frac{(\lambda^q)_{i,j}^{k+1} - (\lambda^q)_{i-1,j}^{k+1}}{\Delta x} \right) + r_u (\lambda^u)_{i,j}^{k+1} = 0,$$

$$-\frac{(\lambda^v)_{i,j}^{k+1} - (\lambda^v)_{i,j}^k}{\Delta t} - f \left(\frac{(\lambda^u)_{i+1,j}^{k+1} + (\lambda^u)_{i,j}^{k+1} + (\lambda^u)_{i+1,j-1}^{k+1} + (\lambda^u)_{i,j-1}^{k+1}}{4} \right)$$

$$- H \left(\frac{(\lambda^q)_{i,j}^{k+1} - (\lambda^q)_{i,j-1}^{k+1}}{\Delta y} \right) + r_v (\lambda^v)_{i,j}^{k+1} = 0,$$

$$-\frac{(\lambda^q)_{i,j}^{k+1} - (\lambda^q)_{i,j}^k}{\Delta t} - g \left(\frac{(\lambda^u)_{i+1,j}^k - (\lambda^u)_{i,j}^k}{\Delta x} + \frac{(\lambda^v)_{i,j+1}^k - (\lambda^v)_{i,j}^k}{\Delta y} \right)$$

$$+ r_q (\lambda^q)_{i,j}^{k+1} = -w \sum_{m=1}^{M} \frac{\delta_{i,i_m} \delta_{j,j_m} \delta_{k,k_m}}{\Delta x \Delta y \Delta t} \left(\hat{q}_{i_m,j_m}^k - d_m \right),$$

where $k = 1, NK - 1$.

$$\frac{(\lambda^u)_{i,j}^{NK}}{\Delta t} = 0,$$

$$\frac{(\lambda^v)_{i,j}^{NK}}{\Delta t} = 0,$$

$$\frac{(\lambda^q)_{i,j}^{NK}}{\Delta t} - g \left(\frac{(\lambda^u)_{i+1,j}^{NK} - (\lambda^u)_{i,j}^{NK}}{\Delta x} + \frac{(\lambda^v)_{i,j+1}^{NK} - (\lambda^v)_{i,j}^{NK}}{\Delta y} \right)$$

$$= -w \sum_{m=1}^{M} \frac{\delta_{i,i_m} \delta_{j,j_m} \delta_{NK,k_m}}{\Delta x \Delta y \Delta t} \left(\hat{q}_{i_m,j_m}^{NK} - d_m \right).$$

$$(\lambda^v)_{i,1}^k = 0,$$

$$(\lambda^v)_{i,NJ}^k = 0 \quad \text{(computational boundary conditions)}.$$

$$(\lambda^u)^k_{0,j} = (\lambda^u)^k_{NI,j},$$

$$(\lambda^u)^k_{NI+1,j} = (\lambda^u)^k_{1,j},$$

$$(\lambda^v)^k_{0,j} = (\lambda^v)^k_{NI,j},$$

$$(\lambda^v)^k_{NI+1,j} = (\lambda^v)^k_{1,j},$$

$$(\lambda^q)^k_{0,j} = (\lambda^q)^k_{NI,j},$$

$$(\lambda^q)^k_{NI+1,j} = (\lambda^q)^k_{1,j}.$$

$$\frac{\hat{u}^{k+1}_{i,j} - \hat{u}^k_{i,j}}{\Delta t} - f\left(\frac{\hat{v}^k_{i,j+1} + \hat{v}^k_{i,j} + \hat{v}^k_{i-1,j+1} + \hat{v}^k_{i-1,j}}{4}\right) + g\left(\frac{\hat{q}^{k+1}_{i,j} - \hat{q}^{k+1}_{i-1,j}}{\Delta x}\right)$$

$$+ r_u \hat{u}^k_{i,j} = (F_u)^k_{i,j} + \left[W^u_f\right]^{-1} (\lambda^u)^{k+1}_{i,j},$$

$$\frac{\hat{v}^{k+1}_{i,j} - \hat{v}^k_{i,j}}{\Delta t} + f\left(\frac{\hat{u}^k_{i+1,j} + \hat{u}^k_{i,j} + \hat{u}^k_{i+1,j-1} + \hat{u}^k_{i,j-1}}{4}\right) + g\left(\frac{\hat{q}^{k+1}_{i,j} - \hat{q}^{k+1}_{i,j-1}}{\Delta y}\right)$$

$$+ r_v \hat{v}^k_{i,j} = (F_v)^k_{i,j} + \left[W^v_f\right]^{-1} (\lambda^v)^{k+1}_{i,j},$$

$$\frac{\hat{q}^{k+1}_{i,j} - \hat{q}^k_{i,j}}{\Delta t} + H\left(\frac{\hat{u}^k_{i+1,j} - \hat{u}^k_{i,j}}{\Delta x} + \frac{\hat{v}^k_{i,j+1} - \hat{v}^k_{i,j}}{\Delta y}\right) + r_q \hat{q}^k_{i,j} = \left[W^q_f\right]^{-1} (\lambda^q)^{k+1}_{i,j}.$$

$$W^u_i\left(\frac{\hat{u}^0_{i,j} - (I^u)_{i,j}}{\Delta t}\right) = \frac{(\lambda^u)^1_{i,j}}{\Delta t} - f\left(\frac{(\lambda^v)^1_{i,j+1} + (\lambda^v)^1_{i,j} + (\lambda^v)^1_{i-1,j+1} + (\lambda^v)^1_{i-1,j}}{4}\right)$$

$$+ H\left(\frac{(\lambda^q)^1_{i,j} - (\lambda^q)^1_{i-1,j}}{\Delta x}\right),$$

$$W^v_i\left(\frac{\hat{v}^0_{i,j} - (I^v)_{i,j}}{\Delta t}\right) = \frac{(\lambda^v)^1_{i,j}}{\Delta t} + f\left(\frac{(\lambda^u)^1_{i+1,j} + (\lambda^u)^1_{i,j} + (\lambda^u)^1_{i+1,j-1} + (\lambda^u)^1_{i,j-1}}{4}\right)$$

$$+ H\left(\frac{(\lambda^q)^1_{i,j} - (\lambda^q)^1_{i,j-1}}{\Delta y}\right) - r_v(\lambda^v)^1_{i,j},$$

$$W^q_i\left(\frac{\hat{q}^0_{i,j} - (I^q)_{i,j}}{\Delta t}\right) = \frac{(\lambda^q)^1_{i,j}}{\Delta t} - r_q(\lambda^q)^1_{i,j}.$$

$$\hat{v}^k_{i,1} = \left[W^v_b\right]^{-1} \Delta y \left\{f\left(\frac{(\lambda^u)^k_{i,1} + (\lambda^u)^k_{i+1,1}}{4}\right) + H\frac{(\lambda^q)^{k+1}_{i,1}}{\Delta y}\right\},$$

$$\hat{v}^k_{i,NJ} = \left[W^v_b\right]^{-1} \Delta y \left\{f\left(\frac{(\lambda^u)^k_{i,NJ-1} + (\lambda^u)^k_{i+1,NJ-1}}{4}\right) - H\frac{(\lambda^q)^{k+1}_{i,NJ-1}}{\Delta y}\right\}.$$

$$\hat{u}^k_{0,j} = \hat{u}^k_{NI,j},$$

$$\hat{u}^k_{NI+1,j} = \hat{u}^k_{1,j},$$

$$\hat{v}_{0,j}^k = \hat{v}_{NI,j}^k,$$
$$\hat{v}_{NI+1,j}^k = \hat{v}_{1,j}^k,$$
$$\hat{q}_{0,j}^k = \hat{q}_{NI,j}^k,$$
$$\hat{q}_{NI+1,j}^k = \hat{q}_{1,j}^k.$$

A.4   Representer calculation

A.4.1   Preamble

The first part of the exercise is to calculate some representers, assemble the representer matrix and verify its algebraic properties. The second part is to find the extremum of a penalty functional by solving the EL equations.

A.4.2   Representer vector

Calculate the representer vector $\mathbf{r}(x, y, t)$.
>   Source code: rep.f
>   To compile the source code: f77-03 rep.f -o rep

A.4.3   Representer matrix

Construct the representer matrix $\mathbf{R}$.
>   Check: Is $\mathbf{R}$ symmetric and positive-definite?

A.4.4   Extremum

Find the extremum of the penalty functional $\mathcal{J}$.

Note: Avoid storing $\mathbf{r}(x, y, t)$ when assembling (3.24).
>   Instead, substitute for the coupling, integrate the backward EL equations and then integrate the forward equations (see §3.1: Accelerating the representer calculation).
>   Check: Is $\hat{\beta}_m = -w\,[\hat{q}(x_m, y_m, t_m) - d_m]$?

A.4.5   Weights

The following values are suggested:

Dynamical weight (u)

$$W_f^u \Delta x\, \Delta y\, \Delta t = (0.25|F_u|)^{-2} \text{ s}^4\text{m}^{-2}.$$

Dynamical weight (v)

$$W_f^v \Delta x \Delta y \Delta t = (0.25|F_u|)^{-2}\ \mathrm{s^4 m^{-2}}.$$

Dynamical weight (q)

$$W_f^q \Delta x \Delta y \Delta t = \infty.$$

Initial weight (u)

$$W_i^u \Delta x \Delta y = \infty.$$

Initial weight (v)

$$W_i^v \Delta x \Delta y = \infty.$$

Initial weight (q)

$$W_i^q \Delta x \Delta y = \infty.$$

Boundary weight (v)

$$W_b^v \Delta x \Delta t = \infty.$$

Data weight (q)

$$w = [0.1\ \max(q_F)]^{-2}\ \mathrm{m^{-2}}.$$

A.5 More representer calculations

A.5.1 Pseudocode, preconditioned conjugate gradient solver

Find the local extremum of the penalty functional $\mathcal{J}$ iteratively (see §3.1, Accelerating the representer calculation). Below is the "pseudocode" of a preconditioned conjugate gradient method (Golub & Van Loan, 1989) for solving

$$\mathbf{U}^{-1/2}(\mathbf{R} + w^{-1}\mathbf{I})\mathbf{U}^{-1/2}\mathbf{U}^{1/2}\hat{\beta} = \mathbf{U}^{-1/2}\mathbf{h},$$

where $\mathbf{U}$ is a preconditioner.

$l = 0$; $\hat{\beta}_0 = 0$; $\mathbf{e}_0 = \mathbf{h}$; $\omega_0 = \| \mathbf{e}_0 \|_2 / \| \mathbf{h} \|_2$
while $\omega_l < \omega_{sc}$
  solve $\mathbf{z}_l = \mathbf{U}^{-1}\mathbf{e}_l$
  $l = l + 1$
  if $l = 1$
  $\mathbf{p}_1 = \mathbf{z}_0$
  else
  $\gamma_l = \mathbf{e}^{\mathrm{T}}{}_{l-1}\mathbf{z}_{l-1} / \mathbf{e}^{\mathrm{T}}{}_{l-2}\mathbf{z}_{l-2}$
  $\mathbf{p}_l = \mathbf{z}_{l-1} + \gamma_l \mathbf{p}_{l-1}$
  endif
  $\alpha_l = \mathbf{e}^{\mathrm{T}}{}_{l-1}\mathbf{z}_{l-1} / \mathbf{p}^{\mathrm{T}}{}_l(\mathbf{R} + w^{-1}\mathbf{I})\mathbf{p}_l$
  $\hat{\beta}_l = \hat{\beta}_{l-1} + \alpha_l \mathbf{p}_l$

$$\mathbf{e}_l = \mathbf{e}_{l-1} - \alpha_l(\mathbf{R} + w^{-1}\mathbf{I})\mathbf{p}_l$$
$$\omega_l = \parallel \mathbf{e}_l \parallel_2 / \parallel \mathbf{h} \parallel_2$$

endwhile

$$\hat{\beta} = \hat{\beta}_l$$

Source code: cgm.f (conjugate gradient solver).

A.5.2   Convolutions for covarying errors

Use "non-diagonal" covariances for initial and dynamical errors (see §2.6, Smoothing norms, covariances and convolutions).

Appendix B

Euler–Lagrange equations for a numerical weather prediction model

The dynamics are those of the standard σ-coordinate, Primitive-Equation model of a moist atmosphere on the sphere (Haltiner and Williams, 1980, p. 17). A penalty functional and the associated Euler–Lagrange equations are given in continuous form; CMFortran code for finite-difference forms is available at an anonymous `ftp` *site. Details of the measurement functionals for reprocessed cloud-track wind observations (see §5.4), and the associated impulses in the adjoint equations, have been suppressed here. The details may be found in the code.*

B.1   Symbols

a_0, Ω	earth's radius, rotation rate
λ, ϕ, σ, t	longitude, latitude, sigma, time
$u, v, \dot{\sigma}$	zonal, meridional, vertical velocity components
$f \equiv 2\Omega \sin \phi$	Coriolis parameter
$\Phi, T, T_{\mathrm{v}}, p, p_* = p/\sigma$	geopotential, temperature, virtual temperature, pressure, surface pressure
q, ρ	specific humidity, density
$l \equiv \ln p_*$	log surface pressure
$R_{\mathrm{d}}, C_{\mathrm{pd}}$	gas constant, specific heat at constant pressure (both for dry air)
$R_{\mathrm{v}}, C_{\mathrm{pv}}$	gas constant, specific heat at constant pressure (both for water vapor)
$\epsilon = R_{\mathrm{v}}/R_{\mathrm{d}}, \ \delta = C_{\mathrm{pv}}/C_{\mathrm{pd}}$	

S_u, S_v, S_T, S_q, S_l	prior estimates of sources of zonal momentum, meridional momentum, heat, humidity, mass
$\langle u \rangle \equiv \int_0^1 u d\sigma,\ u' \equiv u - \langle u \rangle$	vertical average, fluctuation
$\rho_u, \rho_v, \rho_T, \rho_q, \rho_l, \rho_\sigma$	dynamical residuals, or errors in prior estimates of sources
$\mathbf{W}_u, W_T, W_q, W_\Phi, W_l, W_\sigma$	weights for residuals (inverses of prior estimates of covariances of dynamical residuals)
$\mathbf{Q}_u, Q_T, Q_q, Q_\Phi, Q_l, Q_\sigma$	prior estimates of covariances of dynamical residuals
$\mu, \theta, \kappa, \chi, \xi, \omega$	weighted residuals, or adjoint variables
$\mathcal{J} = \mathcal{J}[\mathbf{u}, \dot\sigma, \Phi, T, q, l]$	penalty functional, or estimator for residuals
$F \equiv \partial T / \partial T_v = 1 + (\epsilon^{-1} - 1)q$	moisture factor

$$\mathcal{D} \equiv \frac{1}{a_0 \cos\theta} u_\lambda + \frac{(v \cos\theta)_\theta}{a_0 \cos\theta} + \frac{\partial\dot\sigma}{\partial\sigma} - \frac{\dot\sigma}{\sigma}$$

	divergence
$[\delta u], [\delta v], [\delta\Phi], [\delta T],$ $[\delta\dot\sigma], [\delta l], [\delta q]$	Euler–Lagrange equations for arbitrary variations of $u, v, \Phi, T, \dot\sigma, l, q$
$\mathbf{u}_I, T_I, q_I, l_I$	prior estimates of initial values for $\mathbf{u}, T, q, l$
$\mathbf{V}_u, V_T, V_q, V_l$	weights for residuals or errors in prior estimates of initial values (inverses of initial error covariances)
$\mathbf{O}_u, O_T, O_q, O_l$	prior estimates of initial error covariances for $\mathbf{u}, T, q, l$
Φ_*, V_*, O_*	orography, weight, error covariance
$\bullet, \circ$	four-dimensional and three-dimensional inner products
$r_u^n, r_v^n, r_T^n, r_\Phi^n, r_q^n, r_l^n, r_{\dot\sigma}^n$	representers for n^{th} iterate of Euler–Lagrange equations
$a_u^n, a_v^n, a_T^n, a_\phi^n, a_q^n, a_l^n, a_{\dot\sigma}^n$	adjoint representers

B.2   Primitive Equations and penalty functional

$$u_t + \frac{u}{a_0 \cos\phi} u_\lambda + \frac{v}{a_0} u_\phi + \dot\sigma u_\sigma - \left(f + \frac{u}{a_0}\tan\phi\right)v$$
$$+ \frac{1}{a_0 \cos\phi}(\Phi_\lambda + R_d T_v l_\lambda) - S_u \equiv \rho_u \tag{B.2.1}$$

$$v_t + \frac{u}{a_0 \cos\phi} v_\lambda + \frac{v}{a_0} v_\phi + \dot\sigma v_\sigma + \left(f + \frac{u}{a_0}\tan\phi\right)u$$
$$+ \frac{1}{a_0}(\Phi_\phi + R_d T_v l_\phi) - S_v \equiv \rho_v \tag{B.2.2}$$

$$T_t + \frac{u}{a_0 \cos \phi} T_\lambda + \frac{v}{a_0} T_\phi + \dot{\sigma} T_\sigma$$

$$+ \frac{R_d T_v}{C_{pd} B} \left[\frac{1}{a_0 \cos \phi} u_\lambda + \frac{(v \cos \phi)_\phi}{a_0 \cos \phi} + \sigma \left(\frac{\dot{\sigma}}{\sigma} \right)_\sigma \right] - S_T \equiv \rho_T \quad \text{(B.2.3)}$$

$$q_t + \frac{u}{a_0 \cos \phi} q_\lambda + \frac{v}{a_0} q_\phi + \dot{\sigma} q_\sigma - S_q \equiv \rho_q$$

$$\frac{\partial \Phi}{\partial \ln \sigma} + R_d T_v \equiv \rho_\Phi \quad \text{(B.2.4)}$$

$$l_t + \frac{\langle u \rangle}{a_0 \cos \phi} l_\lambda + \frac{\langle v \rangle}{a_0} l_\phi + \left[\frac{1}{a_0 \cos \phi} \langle u \rangle_\lambda + \frac{(\langle v \rangle \cos \phi)_\phi}{a_0 \cos \phi} \right] - \langle S_l \rangle \equiv \rho_l \quad \text{(B.2.5)}$$

$$\frac{u'}{a_0 \cos \phi} l_\lambda + \frac{v'}{a_0} l_\phi + \left[\frac{1}{a_0 \cos \phi} u'_\lambda + \frac{(v' \cos \phi)_\phi}{a_0 \cos \phi} \right] + \dot{\sigma}_\sigma - S'_l \equiv \rho_\sigma \quad \text{(B.2.6)}$$

$$T_v \equiv T[1 + (\varepsilon^{-1} - 1)q], \quad p = \rho R_d T_v, \quad B \equiv 1 + (\delta - 1)q. \quad \text{(B.2.7)}$$

Weighted residuals

$$\mu \equiv \mathbf{W}_u \bullet \rho_u, \quad \theta \equiv W_T \bullet \rho_T, \quad \kappa \equiv W_q \bullet \rho_q, \quad \text{(B.2.8)}$$

$$\chi = W_\Phi \bullet \rho_\Phi, \quad \xi = W_l \bullet \rho_l, \quad \omega = W_\sigma \bullet \rho_\sigma. \quad \text{(B.2.9)}$$

Penalty functional

$$\mathcal{J} = \mathcal{J}[\mathbf{u}, \dot{\sigma}, \Phi, T, q, l] = \rho_u^* \bullet \mathbf{W}_u \bullet \rho_u + \rho_T \bullet W_T \bullet \rho_T + \rho_q \bullet W_q \bullet \rho_q$$

$$+ \rho_\Phi \bullet W_\Phi \bullet \rho_\Phi + \rho_l \bullet W_l \bullet \rho_l + \rho_\sigma \bullet W_\sigma \bullet \rho_\sigma$$

$$+ \text{(boundary penalties @ } \sigma = 0, 1) + \text{(initial penalties for } \mathbf{u}, T, q, \& l)$$
$$+ \text{(data penalties)}. \quad \text{(B.2.10)}$$

B.3   Euler–Lagrange equations

Note: The symbols $[\delta u]$, etc., indicate that the following equation is the extremal condition for the penalty functional $\mathcal{J}$, with respect to variations of δu, etc.

$[\delta u]$

$$-\mu_t + \frac{u_\lambda \mu}{a_0 \cos \phi} - \frac{(u \mu)_\lambda}{a_0 \cos \phi} - \frac{(v \mu \cos \phi)_\phi}{a_0 \cos \phi} - (\dot{\sigma} \mu)_\sigma - \frac{\tan \phi}{a_0} v \mu + \frac{v_\lambda v}{a_0 \cos \phi}$$

$$+ \left(f + \frac{u \tan \phi}{a_0} \right) v + \frac{\tan \phi}{a_0} uv$$

$$+ \frac{T_\lambda \theta}{a_0 \cos \phi} - \frac{R_d}{C_{pd}} \left(\frac{T_v \theta}{B} \right)_\lambda \frac{1}{a_0 \cos \phi} + \frac{q_\lambda \kappa}{a_0 \cos \phi} + \frac{l_\lambda \xi}{a_0 \cos \phi} - \frac{\xi_\lambda}{a_0 \cos \phi}$$

$$+ \frac{l_\lambda (\omega - \langle \omega \rangle)}{a_0 \cos \phi} - \frac{(\omega_\lambda - \langle \omega \rangle_\lambda)}{a_0 \cos \phi} - \text{(impulses)} = 0 \quad \text{(B.3.1)}$$

$[\delta v]$

$$-v_t + \frac{u_\phi \mu}{a_0} - \left(f + \frac{u \tan \phi}{a_0}\right)\mu - \frac{(uv)_\lambda}{a_0 \cos \phi} + \frac{v_\phi v}{a_0} - \frac{(vv \cos \phi)_\phi}{a_0 \cos \phi} - (\dot\sigma v)_\sigma$$

$$+ \frac{T_\phi \theta}{a_0} - \frac{R_d}{C_{pd}}\left(\frac{T_v \theta}{B}\right)_\phi \frac{1}{a_0} + \frac{q_\phi \kappa}{a_0} + \frac{l_\phi \xi}{a_0} - \frac{\xi_\phi}{a_0} + \frac{l_\phi(\omega - \langle \omega \rangle)}{a_0}$$

$$- \frac{(\omega_\phi - \langle \omega \rangle_\phi)}{a_0} - (\text{impulses}) = 0 \qquad\qquad (B.3.2)$$

$[\delta \Phi]$

$$-\frac{\mu_\lambda}{a_0 \cos \phi} - \frac{(v \cos \phi)_\phi}{a_0 \cos \phi} - (\sigma \chi)_\sigma - (\text{impulses}) = 0 \qquad (B.3.3)$$

$[\delta T]$

$$-\theta_t + \frac{R_d F l_\lambda \mu}{a_0 \cos \phi} + \frac{R_d F l_\phi v}{a_0} - \frac{(u\theta)_\lambda}{a_0 \cos \phi} - \frac{(v\theta \cos \phi)_\phi}{a_0 \cos \phi} - (\dot\sigma \theta)_\sigma + \frac{R_d F D\theta}{C_{pd} B}$$

$$+ R_d F \chi - (\text{impulses}) = 0 \qquad\qquad (B.3.4)$$

$[\delta \dot\sigma]$

$$u_\sigma \mu + v_\sigma v + T_\sigma \theta - \frac{R_d}{C_{pd}}\left(\frac{T_v \theta}{B}\right)_\sigma - \frac{R_d}{C_{pd}}\frac{T_v \theta}{B\sigma} + q_\sigma \kappa - \omega_\sigma - (\text{impulses}) = 0$$

$$(B.3.5)$$

$[\delta l]$

$$-\xi_t - \frac{R_d \langle (T_v \mu)_\lambda \rangle}{a_0 \cos \phi} - \frac{R_d \langle (T_v v \cos \phi)_\phi \rangle}{a_0 \cos \phi} - \frac{(\langle u \rangle \xi)_\lambda}{a_0 \cos \phi} - \frac{(\langle v \rangle \xi \cos \phi)_\phi}{a_0 \cos \phi}$$

$$- \frac{\langle u'\omega \rangle_\lambda}{a_0 \cos \phi} - \frac{\langle v'\omega \cos \phi \rangle_\phi}{a_0 \cos \phi} - (\text{impulses}) = 0 \qquad (B.3.6)$$

$[\delta q]$

$$-\kappa_t + \frac{R_d T F_q l_\lambda \mu}{a_0 \cos \phi} + \frac{R_d T F_q l_\phi v}{a_0} + \frac{R_d T}{C_{pd}} F_q \frac{D\theta}{B} - \frac{R_d}{C_{pd}}\frac{D\theta(\delta - 1)T_v}{B^2}$$

$$- \frac{(u\kappa)_\lambda}{a_0 \cos \phi} - \frac{(v\kappa \cos \phi)_\phi}{a_0 \cos \phi} + R_d F_q \chi - (\dot\sigma \kappa)_\sigma - (\text{impulses}) = 0 \quad (B.3.7)$$

Note: (B.3.5) is a first-order, ordinary differential equation for ω as a function of σ. Solutions are indeterminate, since there is no boundary condition for ω at $\sigma = 0$, nor at $\sigma = 1$. However, the other Euler–Lagrange equations only involve ω' or $\langle u'\omega \rangle$, where $\omega' \equiv \omega - \langle \omega \rangle$, etc., thus the indeterminacy has no effect upon them. The residual ρ_σ in (B.2.6) must satisfy $\langle \rho_\sigma \rangle = 0$; thus its covariance Q_σ must have vanishing integrals with respect to both vertical arguments, and any hypothesis for Q_σ must conform to this requirement. Hence the optimal estimate $\rho_\sigma \equiv Q_\sigma \bullet \omega$ is also unaffected by the indeterminacy, and the vertical integral of the estimate vanishes.

Initial conditions

$$@\ t = 0: \quad \mathbf{u} \cong \mathbf{u}_I, \quad T \cong T_I, \quad q \cong q_I, \quad l \cong l_I. \tag{B.3.8}$$

Contribution to penalty functional

$$\mathcal{J}_I = (\mathbf{u} - \mathbf{u}_I)^* \circ \mathbf{V}_u \circ (\mathbf{u} - \mathbf{u}_I) + (T - T_I) \circ V_T \circ (T - T_I)$$
$$+ (q - q_I) \circ V_q \circ (q - q_I) + (l - l_I) \circ V_l \circ (l - l_I). \tag{B.3.9}$$

Hence

$$@\ t = T: \quad \boldsymbol{\mu} = \mathbf{0}, \quad \theta = 0, \quad \kappa = 0, \quad \xi = 0 \tag{B.3.10}$$

$$@t = 0: \quad -\boldsymbol{\mu} + \mathbf{V}_u \circ (\mathbf{u} - \mathbf{u}_I) = 0, \quad -\theta + V_T \circ (T - T_I) = 0,$$
$$-\kappa + V_q \circ (q - q_I) = 0, \quad -\xi + V_l \circ (l - l_I) = 0, \tag{B.3.11}$$

i.e., $\mathbf{u} = \mathbf{u}_I + \mathbf{O}_u \circ \boldsymbol{\mu}, \quad T = T_I + O_T \circ \theta, \quad q = q_I + O_q \circ \kappa, \quad l = l_I + O_l \circ \xi,$

$$\text{where } \mathbf{O}_{u_{(12)}} \circ \mathbf{V}_{u_{(23)}} = \delta(\mathbf{x}_1 - \mathbf{x}_3)\mathbf{I}, \text{ etc.} \tag{B.3.12}$$

Boundary conditions

$$@\ \sigma = 0, 1: \quad \dot{\sigma} = 0 \tag{B.3.13}$$

$$@\ \sigma = 1: \quad \Phi \cong \Phi_*. \tag{B.3.14}$$

Contribution to penalty functional

$$\mathcal{J}_* = (\Phi - \Phi_*) \circ V_* \circ (\Phi - \Phi_*). \tag{B.3.15}$$

Hence

$$@\ \sigma = 0: \quad \chi = 0, \tag{B.3.16}$$

$$@\ \sigma = 1: \quad \Phi = \Phi_* - O_* \circ \chi. \tag{B.3.17}$$

B.4   Linearized Primitive Equations

Note: The labels (LPE1) etc. refer to lines of the CMFortran code for the finite-difference equations; the code is available at an anonymous ftp site (ftp.oce.orst.edu, /dist/chua/IOM/IOSU).

$$u_t^n + \frac{u^{n-1}u_\lambda^n}{a_0 \cos \phi} + \frac{v^{n-1}u_\phi^n}{a_0} + \dot{\sigma}^{n-1}u_\sigma^n - \left(f + \frac{u^{n-1}\tan \phi}{a_0} \right) v^n$$

$$+ \frac{1}{a_0 \cos \phi} \left(\Phi_\lambda^n + R_d T_v^{n-1} l_\lambda^n \right) - S_u^n \equiv \rho_u^n. \tag{LPE1}\ \ (\text{B.4.1})$$

$$v_t^n + \frac{u^{n-1}v_\lambda^n}{a_0 \cos\phi} + \frac{v^{n-1}v_\phi^n}{a_0} + \dot{\sigma}^{n-1}v_\sigma^n + \left(f + \frac{u^{n-1}\tan\phi}{a_0}\right)u^n$$

$$+ \frac{1}{a_0}\left(\Phi_\phi^n + R_d T_v^{n-1}l_\phi^n\right) - S_v^n \equiv \rho_v^n. \qquad \text{(LPE2) (B.4.2)}$$

$$T_t^n + \frac{u^{n-1}T_\lambda^n}{a_0 \cos\phi} + \frac{v^{n-1}T_\phi^n}{a_0} + \dot{\sigma}^n\, T_\sigma^{n-1}$$

$$+ \frac{R_d T_v^{n-1}}{C_{pd}B^{n-1}}\left[\frac{u_\lambda^n}{a_0 \cos\phi} + \frac{(v_\phi^n \cos\phi)_\phi}{a_0 \cos\phi} + \frac{\partial \dot{\sigma}^n}{\partial \sigma} - \frac{\dot{\sigma}^n}{\sigma}\right] - S_T^n \equiv \rho_T^n.$$

$$\text{(LPE3) (B.4.3)}$$

$$q_t^n + \frac{u^{n-1}q_\lambda^n}{a_0 \cos\phi} + \frac{v^{n-1}q_\phi^n}{a_0} + \dot{\sigma}^{n-1}q_\sigma^n - S_q^n \equiv \rho_q^n. \qquad \text{(LPE4) (B.4.4)}$$

$$\frac{\partial \Phi^n}{\partial \ln\sigma} + R_d T^n F^{n-1} \equiv \rho_\Phi^n. \qquad \text{(LPE5) (B.4.5)}$$

$$l_t^n + \frac{\langle u^{n-1}\rangle l_\lambda^n}{a_0 \cos\phi} + \frac{\langle v^{n-1}\rangle l_\phi^n}{a_0} + \left[\frac{1}{a_0 \cos\phi}\langle u^n\rangle_\lambda + \frac{(\langle v^n\rangle \cos\phi)_\phi}{a_0 \cos\phi}\right] - \langle S_l^n\rangle \equiv \rho_l^n.$$

$$\text{(LPE6) (B.4.6)}$$

$$\frac{u'^{n-1}l_\lambda^n}{a_0 \cos\phi} + \frac{v'^{n-1}l_\phi^n}{a_0} + \left[\frac{1}{a_0 \cos\phi}u_\lambda'^n + \frac{(v'^n \cos\phi)_\phi}{a_0 \cos\phi}\right] + \frac{\partial \dot{\sigma}^n}{\partial \sigma} - S_l'^{n-1} \equiv \rho_{\dot\sigma}^n.$$

$$\text{(LPE7) (B.4.7)}$$

$$T_v^n = T^n[1 + (\varepsilon^{-1} - 1)q^n]. \qquad \text{(B.4.8)}$$

B.5   Linearized Euler–Lagrange equations

$[\delta u^n]$

$$-\mu_t^n - \frac{(u^{n-1}\mu^n)_\lambda}{a_0 \cos\phi} - \frac{(v^{n-1}\mu^n \cos\phi)_\phi}{a_0 \cos\phi} - (\dot{\sigma}^{n-1}\mu^n)_\sigma + \left(f + \frac{u^{n-1}\tan\phi}{a_0}\right)v^n$$

$$- \frac{R_d}{C_{pd}}\frac{1}{a_0 \cos\phi}\left(\frac{T_v^{n-1}\theta^n}{B^{n-1}}\right)_\lambda - \frac{\xi_\lambda^n}{a_0 \cos\phi} - \frac{(\omega_\lambda^n - \langle\omega^n\rangle_\lambda)}{a_0 \cos\phi}$$

$$= -\frac{u_\lambda^{n-1}\mu^{n-1}}{a_0 \cos\phi} + \frac{v^{n-1}\tan\phi}{a_0}\mu^{n-1} - \frac{v_\lambda^{n-1}v^{n-1}}{a_0 \cos\phi} - \frac{\tan\phi}{a_0}u^{n-1}v^{n-1}$$

$$- \frac{T_\lambda^{n-1}\theta^{n-1}}{a_0 \cos\phi} - \frac{q_\lambda^{n-1}\kappa^{n-1}}{a_0 \cos\phi} - \frac{l_\lambda^{n-1}\xi^{n-1}}{a_0 \cos\phi}$$

$$- l_\lambda^{n-1}\frac{(\omega^{n-1} - \langle\omega^{n-1}\rangle)}{a_0 \cos\phi} + (\text{impulses})^n. \qquad \text{(LELE1) (B.5.1)}$$

$[\delta v^n]$

$$
- v_t^n - \frac{(u^{n-1} v^n)_\lambda}{a_0 \cos \phi} - \frac{(v^{n-1} v^n \cos \phi)_\phi}{a_0 \cos \phi} - (\dot{\sigma}^{n-1} v^n)_\sigma - \left(f + \frac{u^{n-1} \tan \phi}{a_0} \right) \mu^n
$$

$$
- \frac{R_d}{C_{pd}} \left(\frac{T_v^{n-1} \theta^n}{B^{n-1}} \right)_\phi \frac{1}{a_0} - \frac{\xi_\phi^n}{a_0} - \frac{(\omega_\phi^n - \langle \omega_\phi^n \rangle)}{a_0}
$$

$$
= - \frac{u_\phi^{n-1} \mu^{n-1}}{a_0} - \frac{v_\phi^{n-1} v^{n-1}}{a_0 \cos \phi} - \frac{T_\phi^{n-1} \theta^{n-1}}{a_0} - \frac{q_\phi^{n-1} \kappa^{n-1}}{a_0} - \frac{l_\phi^{n-1} \xi^{n-1}}{a_0}
$$

$$
- \frac{l_\phi^{n-1}}{a_0} (\omega^{n-1} - \langle \omega^{n-1} \rangle) + (\text{impulses})^n. \qquad \text{(LELE2) (B.5.2)}
$$

$[\delta \Phi^n]$

$$
\frac{-\mu_\lambda^n}{a_0 \cos \phi} - \frac{(v^n \cos \phi)_\lambda}{a_0 \cos \phi} - (\sigma \chi)_\sigma^n = (\text{impulses})^n. \qquad \text{(LELE3) (B.5.3)}
$$

$[\delta T^n]$

$$
- \theta_n^t - \frac{(u^{n-1} \theta^n)_\lambda}{a_0 \cos \phi} - \frac{(v^{n-1} \theta^n \cos \phi)_\phi}{a_0 \cos \phi} + R_d F^{n-1} \chi^n
$$

$$
= - \frac{R_d F^{n-1} l_\lambda^{n-1} \mu^{n-1}}{a_0 \cos \phi} - \frac{R_d F^{n-1} l_\phi^{n-1} v^{n-1}}{a_0} - \frac{R_d F^{n-1} \mathcal{D}^{n-1} \theta^{n-1}}{c_{pd} B^{n-1}}
$$

$$
+ \left(\dot{\sigma}^{n-1} \theta^{n-1} \right)_\sigma + (\text{impulses})^n. \qquad \text{(LELE4) (B.5.4)}
$$

$[\delta \dot{\sigma}^n]$

$$
T_\sigma^{n-1} \theta^n - \frac{R_d}{C_{pd}} \left(\frac{T_v^{n-1} \theta^n}{B^{n-1}} \right)_\sigma - \frac{R_d}{C_{pd}} \frac{T_v^{n-1} \theta^n}{B^{n-1} \sigma} - \omega_\sigma^n
$$

$$
= - u_\sigma^{n-1} \mu^{n-1} - v_\sigma^{n-1} v^{n-1} - q_\sigma^{n-1} \kappa^{n-1} + (\text{impulses})^n. \qquad \text{(LELE5) (B.5.5)}
$$

$[\delta l^n]$

$$
- \xi_t^n - \frac{R_d \langle (T_v^{n-1} \mu^n)_\lambda \rangle}{a_0 \cos \phi} - \frac{R_d \langle (T_v^{n-1} v^n \cos \phi)_\phi \rangle}{a_0 \cos \phi} - \frac{(\langle u^{n-1} \rangle \xi^n)_\lambda}{a_0 \cos \phi}
$$

$$
- \frac{(\langle v^{n-1} \rangle \xi^n \cos \phi)_\phi}{a_0 \cos \phi} - \frac{\langle (u'^{m-1} \omega^n)_\lambda \rangle}{a_0 \cos \phi} - \frac{\langle (v'^{n-1} \omega^n \cos \phi)_\phi \rangle}{a_0 \cos \phi}
$$

$$
= (\text{impulses})^n. \qquad \text{(LELE6) (B.5.6)}
$$

$[\delta q^n]$

$$
- \kappa_t^n - \frac{(u^{n-1} \kappa^n)_\lambda}{a_0 \cos \phi} - \frac{(v^{n-1} \kappa^n \cos \phi)_\phi}{a_0 \cos \phi} - (\dot{\sigma}^{n-1} \kappa^n)_\sigma
$$

$$= -\frac{R_d T^{n-1} F_q^{n-1} l_\lambda^{n-1} \mu^{n-1}}{a_0 \cos\phi} - \frac{R_d T^{n-1} F_q^{n-1} l_\phi^{n-1} v^{n-1}}{a_0} - \frac{R_d T^{n-1} F_q^{n-1} \mathcal{D}^{n-1} \theta^{n-1}}{C_{pd} B^{n-1}}$$

$$+ \frac{R_d T_v^{n-1} \mathcal{D}^{n-1} \theta^{n-1} (\delta - 1)}{C_{pd} B^2} - R_d F_q^{n-1} \chi^{n-1} + (\text{impulses})^n. \quad \text{(LELE7) (B.5.7)}$$

B.6   Representer equations

[u]

$$r_{u_t}^n + \frac{u^{n-1} r_{u_\lambda}^n}{a_0 \cos\phi} + \frac{v^{n-1} r_{u_\phi}^n}{a_0} + \dot\sigma^{n-1} r_{u_\sigma}^n - \left(f + \frac{u^{n-1} \tan\phi}{a_0}\right) r_v^n$$

$$+ \frac{1}{a_0 \cos\phi}\left(r_{\Phi_\lambda}^n + R_d T_v^{n-1} r_{l_\lambda}^n\right) = \left(\mathbf{Q}_u \bullet \mathbf{a}_u^n\right)_u. \quad \text{(RE1) (B.6.1)}$$

[v]

$$r_{v_t}^n + \frac{u^{n-1} r_{v_\lambda}^n}{a_0 \cos\phi} + \frac{v^{n-1} r_{v_\phi}^n}{a_0} + \dot\sigma^{n-1} r_{v_\sigma}^n + \left(f + \frac{u^{n-1} \tan\phi}{a_0}\right) r_u^n$$

$$+ \frac{1}{a_0}\left(r_{\Phi_\phi}^n + R_d T_v^{n-1} r_{l_\phi}^n\right) = \left(\mathbf{Q}_u \bullet \mathbf{a}_u^n\right)_v. \quad \text{(RE2) (B.6.2)}$$

[T]

$$r_{T_t}^n + \frac{u^{n-1} r_{T_\lambda}^n}{u_0 \cos\phi} + \frac{v^{n-1} r_{T_\phi}^n}{a_0} + r_{\dot\sigma}^n T_\sigma^{n-1} + \frac{R_d T_v^{n-1}}{C_{pd} B^{n-1}}\left[\frac{r_{u_\lambda}^n}{a_0 \cos\phi} + \frac{(\cos\phi r_v^n)_\phi}{a_0 \cos\phi}\right.$$

$$\left. + \sigma\left(\frac{r_{\dot\sigma}^n}{\sigma}\right)_\sigma\right] = Q_T \bullet a_T^n. \quad \text{(RE3) (B.6.3)}$$

[q]

$$r_{q_t}^n + \frac{u^{n-1} r_{q_\lambda}^n}{a_0 \cos\phi} + \frac{v^{n-1} r_{q_\phi}^n}{a_0} + \dot\sigma^{n-1} r_{q_\sigma}^n = Q_q \bullet a_q^n. \quad \text{(RE4) (B.6.4)}$$

[Φ]

$$\frac{\partial r_\Phi^n}{\partial \ln\sigma} + R_d r_T^n F^{n-1} = Q_\Phi \bullet a_\Phi^n. \quad \text{(RE5) (B.6.5)}$$

[l]

$$r_{l_t}^n + \frac{\langle u^{n-1}\rangle r_{l_\lambda}^n}{a_0 \cos\phi} + \frac{\langle v^{n-1}\rangle r_{l_\phi}^n}{a_0} + \left[\frac{\langle r_u^n\rangle_\lambda}{a_0 \cos\phi} + \frac{\cos\phi \langle r_v^n\rangle_\phi}{a_0 \cos\phi}\right] = Q_l \bullet a_l^n.$$

$$\text{(RE6) (B.6.6)}$$

[σ̇]

$$\frac{u'^{n-1} r_{l_\lambda}^n}{a_0 \cos\phi} + \frac{v'^{n-1} r_{l_\phi}^n}{a_0} + \left[\frac{r_{u_\lambda}'^n}{a_0 \cos\phi} + \frac{(r_v'^n \cos\phi)_\phi}{a_0 \cos\phi}\right] + \frac{\partial r_{\dot\sigma}^n}{\partial \sigma} = Q_{\dot\sigma} \bullet a_{\dot\sigma}^n.$$

$$\text{(RE7) (B.6.7)}$$

B.7   Representer adjoint equations

$[\delta u]$

$$
-a_{u_t}^n - \frac{\left(u^{n-1}a_u^n\right)_\lambda}{a_0\cos\phi} - \frac{\left(v^{n-1}a_u^n\cos\phi\right)_\phi}{a_0\cos\phi} - \left(\dot\sigma^{n-1}a_u^n\right)_\sigma + \left(f + \frac{u^{n-1}\tan\phi}{a_0}\right)a_v^n
$$

$$
- \frac{R_d}{C_{pd}}\left(\frac{T_v^{n-1}a_T^n}{B^{n-1}}\right)_\lambda \frac{1}{a_0\cos\phi} - \frac{a_{l_\lambda}^n}{a_0\cos\phi} - \frac{\left(a_{\dot\sigma_\lambda}^n - \langle a_{\dot\sigma_\lambda}^n\rangle\right)}{a_0\cos\phi} = \text{(impulse)}.
$$

$$
\text{(RAE1) (B.7.1)}
$$

$[\delta v]$

$$
-a_{v_t}^n - \frac{\left(u^{n-1}a_v^n\right)_\lambda}{a_0\cos\phi} - \frac{\left(v^{n-1}a_v^n\cos\phi\right)_\phi}{a_0\cos\phi} - \left(\dot\sigma^{n-1}a_v^n\right)_\sigma - \left(f + \frac{u^{n-1}\tan\phi}{a_0}\right)a_u^n
$$

$$
- \frac{R_d}{C_{pd}}\frac{1}{a_0}\left(\frac{T_v^{n-1}a_T^n}{B^{n-1}}\right)_\phi - \frac{a_{l_\phi}^n}{a_0} - \frac{\left(a_{\dot\sigma_\phi}^n - \langle a_{\dot\sigma_\phi}^n\rangle\right)}{a_0} = \text{(impulse)}.
$$

$$
\text{(RAE2) (B.7.2)}
$$

$[\delta\Phi]$

$$
-\frac{a_{u_\lambda}^n}{a_0\cos\phi} - \frac{\left(a_v^n\cos\phi\right)_\phi}{a_0\cos\phi} - \left(\sigma a_\Phi^n\right)_\sigma = \text{(impulse)}.
$$

$$
\text{(RAE3) (B.7.3)}
$$

$[\delta T]$

$$
-a_{T_t}^n - \frac{\left(u^{n-1}a_T^n\right)_\lambda}{a_0\cos\phi} - \frac{\left(v^{n-1}a_T^n\cos\phi\right)_\phi}{a_0\cos\phi} + R_d F^{n-1}a_\Phi^n = \text{(impulse)}.
$$

$$
\text{(RAE4) (B.7.4)}
$$

$[\delta\dot\sigma]$

$$
T_\sigma^{n-1}a_T^n - \frac{R_d}{C_{pd}}\left(\frac{T_v^{n-1}a_T^n}{B^{n-1}}\right)_\sigma - \frac{R_d}{C_{pd}}\frac{T_v^{n-1}a_T^n}{B^{n-1}\sigma} - \frac{\partial a_{\dot\sigma}^n}{\partial\sigma} = \text{(impulse)}.
$$

$$
\text{(RAE5) (B.7.5)}
$$

$[\delta l]$

$$
-R_d\frac{\langle T_v^{n-1}a_u^n\rangle_\lambda}{a_0\cos\phi} - R_d\frac{\langle T_v^{n-1}a_v^n\cos\phi\rangle_\phi}{a_0\cos\phi} - a_{l_t}^n - \frac{\left(\langle u^{n-1}\rangle a_l^n\right)_\lambda}{a_0\cos\phi} - \frac{\left(\langle v^{n-1}\rangle a_l^n\cos\phi\right)_\phi}{a_0\cos\phi}
$$

$$
- \frac{\langle u'^{n-1}a_{\dot\sigma}^n\rangle_\lambda}{a_0\cos\phi} - \frac{\langle v'^{n-1}a_{\dot\sigma}^n\cos\phi\rangle_\phi}{a_0\cos\phi} = \text{(impulse)}. \qquad \text{(RAE6) (B.7.6)}
$$

$[\delta q]$

$$
-a_{q_t}^n - \frac{\left(u^{n-1}a_q^n\right)_\lambda}{a_0\cos\phi} - \frac{\left(v^{n-1}a_q^n\cos\phi\right)_\phi}{a_0\cos\phi} - \left(\dot\sigma^{n-1}a_q^n\right)_\sigma = \text{(impulse)}.
$$

$$
\text{(RAE7) (B.7.7)}
$$

Author index

Subject index